# THE PROTECTED WILL NEVER KNOW

## DON MEYER

two peas publishing

# COPYRIGHT INFORMATION

**The Protected Will Never Know**
All Rights Reserved © 2003 by Don Meyer
No part of this book may be reproduced or transmitted in any form or by any means, graphic, electronic, or mechanical, including photocopying, recording, taping, or by any information storage retrieval system, without the written permission of the publisher.

**10th Anniversary Edition**
© 2012 by Don Meyer

Book Design by Paula Rozelle Hanback

ISBN 978-1-938271-14-4

Published by:
Two Peas Publishing
PO Box 1193
Franklin, TN U.S.A.

# ALSO BY DON MEYER

*The Sheriff Tom Monason Trilogy:*

*Winter Ghost*

*Mckenzie Affair*

*Uncle Denny*

**Jennifer's Plan**

**The American War**

www.dpmeyer.com

## ACKNOWLEDGEMENTS

My deepest appreciation and gratitude, to my editor Ralph Kimball, for his willingness, to take on this project, with short notice and his ability to turn it around with an even shorter deadline.

A bushel of thanks to Lia Scott Price, an author of extraordinary talent, for support and help in getting this off my desk and into book form.

And in no particular order, my thanks and appreciation to:

The University of Notre Dame for permission to mention their name.

Playboy Magazine for permission to mention the name PLAYBOY.

The Chicago Radio Syndicate for permission to refer to CHICKENMAN, the White Winged Warrior.

Parker Brothers for the use of the trademark MONOPOLY®.

A special thanks to my daughter, Carroll, for her insistence that I get this manuscript out of the box, where it sat for twenty five years, and do something with it.

# FOREWORD

This book is a collection of stories based on the time I spent in Vietnam from 1969 to 1970. Unlike others I was able to "talk about it." In fact I was able to "get over it" as we said back then.

As I was arriving in Vietnam in late November of 1969 a war protest march of about a quarter million people was happening in Washington DC. The charges and subsequent court martial were starting to surface for the Songmy (later know as the My Lai) massacre.

On December 1, 1969 the first draft lottery since WWII was held for troop selection in 1970. I never did find out my "number". Of course it didn't matter. I was already there.

The Vietnam "conflict" was now under the direction of the fifth different US president, twenty-fifth year of US involvement and fifth year of direct US combat operations. Peace talks were underway in Paris (well, sort of).

I was raised on the south side of Chicago, a block and a half from the ballpark by a single parent. My life had a very narrow focus. I grew up in an Italian neighborhood comprised of working class people. I attended a Catholic high school… and promptly went to work the day after graduation. College was not an option for most of us in that neighborhood.

I'd like to say being street smart was probably my strongest educational quality. I knew how to survive, and make the best of any situation. While we and the neighborhood were considered poor or not well off or… To those of us who lived there it was home and it was the best we had.

We and certainly I were not very worldly. We knew what was happening in the "hood" and the city of Chicago and maybe some things about the state, but a war (excuse me "conflict") happening halfway around the world, not a chance. Sure we watched the news, but if it wasn't happening down the street, it just didn't matter.

My draft notice came in the summer of 1969 and I was in Vietnam by Thanksgiving. Pretty much before I even knew where it was I was there.

My outlook on life was rather simple and naïve. It was as traumatic to be away from my surroundings (Chicago, the city, the neighborhood) as it was to be in Vietnam. I could have been anywhere and it would have been just as traumatic. The fact that Vietnam was a war zone was just a minor piece of the picture.

I've always been a positive person, a "glass is half full" kind of guy, so my outlook has always taken on the positive. Sure I could dwell on the "how bad it was" concept, but like anything in life you learn to adapt or you die and in this case, dying was an everyday occurrence.

The original title for this book was "For Those Who Weren't There" because I wrote these stories to let others know what it was like to live, exist and survive in Vietnam on a day to day basis. Being there was more than just staying alive. It was surviving the elements, the army, the bullshit and the knowledge you just didn't matter.

I returned from Vietnam in late 1970, served out the rest of my time, returned home, married my future ex-wife and got back to business.

The actual concept for this book came from an impromptu short story I had to write in a college English class. The time was the early 1970s. I was attending a junior college on the GI bill and trying to get on with my life. The classes were largely made up of Vietnam veterans. Some of us talked, some of us didn't.

From the basis of that handwritten short story, I composed others and eventually had a "bunch" of handwritten notes and stories.

Shortly after my divorce, (Vietnam related? Not in the least!) I was forced to move back in with my mom and stepfather and it was there that I pulled out my mom's old big black manual typewriter and set about putting these notes and stories into some semblance of order.

I had a journal of activities I had kept, and most importantly, a host of letters that I had written to my ex, another girl, my mom and letters I had addressed as "your man on the scene" that I sent back to the folks where I worked. So I had a wealth of information to refer to, but most importantly I had my memories, that were still quite sharp that soon after.

It was enough to fill in the blanks as I set about to put words to paper. The first few drafts were quite atrocious as I pounded on that manual typewriter, but I was able to create a completed manuscript. One of the secretaries at the office I worked at retyped it into a new fangled word processor the office had at the time called a mag-card machine. It created these magnetic cards that could be used to store the original and allow you to make changes and re-print. We are talking the early seventies here. Technology was on the move.

We actually created a clean typewritten manuscript that I sent off to several publishers and was rejected by everyone. By 1979, I needed to get on with my life and threw the manuscript, still in its envelope stamped "Unsolicited material, return to sender" into a box with my notes and stuff and put it on a shelf in a closet. It remained there until I moved in 2002 and was forced to "deal with it" again some twenty-three years later.

I took on the effort to re-key those words and pages into my computer if for no other reason than to have a copy of it stored somewhere. My daughter has been after me for years to do something with it. She has been one of the few people to ever read it all the way through and one of the earlier versions at that.

So, after all these years I'm doing something with it.

[I elected to put out a new edition for a number of reasons, the single most was the size of the print—most of us at this age can hardly see it. This was the very first book I put out using a new publishing process that was coming in to vogue back then. And while I was thrilled to have my book in print, I always felt I wish I had a do over, specifically for the reason stated above and especially now that I know and understand a lot more. In addition I have also added an epilogue to "finish the story" as often requested. However, as before I left this work in the voice of that 25 year old that originally wrote this manuscript, with all its warts and foibles. The original manuscript was actually written in 1977 and 25 years later in 2002, I literally retyped that document word for word from its white bond paper into a computer for easier access. Because I did most of the work and preparation in 2002, I have chosen that year as the rebirth of this journey and while I had every intention to do a fifth anniversary edition, for the simple reason described above that just never materialized. So now here it is ten years later and this time I am following through with that project.]

# ARRIVAL

**THE PLANE SUDDENLY BECAME VERY QUIET.** The flight from Japan was more subdued than the flight from Alaska. That leg had been one continuous party, but now it was different.

We had boarded the plane in the early hours of the morning in San Francisco to start our trek. Our first stop was in Alaska where some of the guys found a bar open at 8:30 in the morning to settle the nerves they had said. The next leg had taken us from Alaska to Japan for a brief stopover. Our next stop was Bien Hoa (I heard pronounced ben wah), a town just north of Saigon. We would land in a few minutes.

As we disembarked, the oppressive heat seemed to hit me most of all. So this is it. This is the place I had wondered about since I found out I would be assigned here. Although I will admit I had regarded it with a mischievously curious ambition to be here, I was scared at the actual aspect of physically being present in country for my tour of duty.

We all seemed to drift about aimlessly in this dark place they called a terminal. All of our duffel bags were dumped in an even

# 2  THE PROTECTED WILL NEVER KNOW

darker place behind the terminal. I can still remember some asshole yelling out.

"Alright you got three minutes to find your bag and get into the next building for your money change over."

All two hundred of us descended on the pile at once. I believe at least two guys got trampled, but like everything else we did in this organization, we managed to get through that mess. Again, remembering back I wondered how long I carried Howard Jordan's duffel bag around that first night (morning)!

I don't remember much about all of the buildings we went through that night (morning). In all of the time I have been in this organization, I don't believe there was ever a time we did anything during normal hours. Actually, the army is a 24/7 operation. There are no normal hours.

Anyway, I do remember the money changeover, mostly because they wouldn't let us keep anything under a nickel. Of course, everyone had four pennies they had to forfeit. Not a bad racket I thought. This one particular guy behind me had been doing some fast calculations. Suddenly he tapped me on the shoulder.

"Do you realize they are taking in about a hundred and a half a week?"

"So?" I answered not having a clue of what he was talking about.

"So… so… the man said we could be here at least a week before we would be reassigned and that puts us to Friday. At which time we could hit this place for the week's receipts and blow the following morning to some assignment somewhere."

I couldn't believe my ears. This guy was actually thinking of robbing all those pennies. As I turned to him, still not believing what I had heard, I casually mentioned to him that it would be somewhat ridiculous to take such a chance for a lousy hundred and fifty dollars in pennies.

His eyes burst open, his face began turning a deep shade of red and yes, I do believe he was starting to foam at the month. I thought, any second, he was going to strike me. The anxiety started to mount. My heart began to race. I waited for his anger to subside. I waited for what seemed like hours, but were only seconds, before he finally spoke.

"Why you stupid fucker, you stupid ignorant fucker. You actually thought I wanted those goddamn pennies?" He paused to regain his composure. "Just take a gander at that sign over there."

As the words slowly sank in, I realized I had really blown it this time. The sign in its entire splendor stated very simply: Pay Center.

I was at a loss for words. I couldn't find anything to say, so he spoke again.

"What did you think I meant by a hundred and a half?

He could tell by my stunned look that I was still not quite with it, so he took the initiative again.

"Look pal if we are coming in and we are switching to this stuff..." He pointed to the funny looking paper they were giving everybody for their American greenbacks. "...then it stands to reason that the dudes that are leaving are going to be switching back."

I could hardly disagree with his reasoning. It was very logical. However, I must have given the impression that I still wasn't with him, because he let out a deep sigh and started to explain again.

"Look, let's try it again. As I said, they would be switching back, so what with all the greenbacks we and all the rest before

## | 4 | THE PROTECTED WILL NEVER KNOW

and after us have turned in this week, plus the large cash reserve they would have on hand for those leaving, I figure that they would probably have about a hundred and fifty thousand on hand Friday night prior to the weekend departures."

"Aha…" I exclaimed. "That's where your theory is all wet." I knew I had him now. I was starting to regain some of my composure. "It stands to reason…" I said, "…that they would use the greenbacks they collected to pay out to those leaving."

Proud of myself now, I had finally regained some of my reasoning. He started to heat up again, but he didn't give it a go this time. He simply shook his head and mumbled.

"Boy, you really are a stupid fucker. That would be the logical thing to do wouldn't it? However this being the army, and we know they never do anything logical, which makes my idea even more practical. Doesn't it?"

We continued to inch closer to the cashier in total silence now. Since my lack of understanding had shattered his dreams, we no longer seemed able to communicate.

It was my turn now to switch my greenbacks for that funny money. As I watched them count out the switch and confiscate my pennies, I couldn't help but wonder. So this is Vietnam, this is what it will be like to fight a war!

# RECEPTION

As the jet roared overhead, another group held their breath in anticipation, hoping this might be their "freedom bird" (commercial jet back to the states) this time.

There is no other way to describe reception station than as a place of desperation. For those that are waiting for that flight to take them back to the world (home... states... anywhere but here, was back to the world) and those waiting to be assigned somewhere, the feeling of anticipation hung heavy in the air.

I watched silently as those who had done their time milled around hoping this time they would be boarding that flight home. After twelve months in country your tour of duty was over and you reached DEROS (or date eligible to return from overseas).

Those going back to the world had that distant look in their eyes along with a certain body language that said what a year here had been like. Deeply tanned, hardened, deliberate movements, no wasted actions. Even though they knew their tour of duty was over there was still that feeling of apprehension until they boarded that freedom bird back to the world.

# | 6 | THE PROTECTED WILL NEVER KNOW

For those of us who were hanging around waiting for reassignment, we more or less preferred to just sort of rot here.

Each day we stood milling about waiting for our name (hell, the army doesn't use names, but however they referred to us) to be called, wondering where we would be sent. I again felt the nausea building inside my gut aging in time to the cadence of names (?) being called off for reassignment, still waiting for mine.

There was a building off to one side that held all the orders, orders to go home, go somewhere, go to hell, go anywhere… but as usual it was "off limits". That big sign that was posted everywhere you might be able to get information. It was used to keep us in our proper places. Boy, how I hated that lousy sign. I had been here three days without a word.

I watched all of my friends and everyone else who had come through with me be reassigned in a day to a day and a half. No one was allowed to ask why you hadn't been reassigned yet, until you were here for three days. I couldn't wait. Well, this was my third day.

I approached the building with the cunning and caution you might use in approaching the VC (Viet Cong). This was probably ridiculous, but I was really afraid of going in there. For all I knew, it was probably controlled by the VC and we were really all prisoners being reassigned to death camps…

Standing in the doorway was a Spec 4 (Specialist fourth class, the army's answer to that rank between private and sergeant) all nice and clean. His uniform was neatly pressed as if he was serving stateside duty somewhere. How could this be I wondered? Here I was in the same clothes I had arrived in three days ago about ready to ferment.

I approached him with my fear having built to a nerve

shattering climax and started to ask him if this was the right place, but he appeared to ignore me, or did he? He stood there, at first not noticing me, then staring blankly at me. I kept speaking but still no response. It was then that I realized that although I thought I was speaking, I was really just moving my mouth without saying anything. My fear of this place had driven me speechless. What was this place that had put me in such a state?

Somehow words escaped from my mouth. I became aware of his directing me to a table in the back with another "spic and span" clerk looking bored out of his mind sitting there with a bunch of files (those precious 201 personal history files) in front of him. I staggered over and sat down. I waited a few minutes for the clerk to speak.

"What can I do for you?"

It rang out as if I was on trial for murder... YOU!

"Ahhhh... I've been here for three days and..."

"So."

"Well, I was under the impression that if you did not get assigned within three days they were to report here and..."

"You've been here four days?"

"No three..."

"Then what are you doing here?"

"As I said..."

"No, as I said, four days. After you have been here three whole days, then on the fourth day you come here to see about travel orders, pay vouchers, supply chits, meal tickets and traveling papers."

"But this is my third day, actually fourth, because I arrived in the middle of the night three days ago and I haven't been reassigned yet. I just thought that I might be able to get some information. By the way, my meal ration ran out last night so I couldn't get breakfast this morning."

My heart was racing again. I was on trial for murder. I had violated a cardinal rule. Don't question the army's logic of just following orders. The clerk appeared to be a little more receptive now.

"Well as long as you are here, I'll see if I can find out anything. What is you name, number etc.?"

"My name?"

"Yes, name, number, something. "

"Ahhhh... yes my name is a..."

"YOU DO know your name?"

I was stalling. I had a way out. If I gave him the wrong name he wouldn't find out anything and I would just walk out, tell him thanks and... But my God what if I did give him a wrong name and he knew it? My heart was still racing. Should I lie and take my chances? I could just drift back in with the rest of the transients. No one would be the wiser. I would be free of this building. But what if I was found out. What would they do this time? I had gotten away with it three nights ago.

"For Christ sakes, what is your fucking name?"

"Howard Jordan."

"Who?"

"Jordan, Howard Jordan."

"Alright wait here, I'll be right back."

My heart was on fire now. I had done it. I had sent him on a false lead. However that wasn't a total lie. Hadn't that been my name most of the time three nights ago? I carried his duffle bag most of the night. It seemed almost fitting that now that I needed help he should help me, whoever he was.

"Jordan..."

The spic and span clerk was looking right at me now. I had been found out. This was it. I had impersonated someone. My skin began to crawl.

"Yes." Was all I could say.

"Be right there."

I waved back at him and smiled. The clerk was talking to someone, an officer. Oh shit this was it, Leavenworth (Federal Penitentiary) for sure. Why… why had I done it? I watched the officer point and the clerk look at me.

He approached the table hurriedly and threw the file he had in his hand down, swearing as he did.

"Those fucking lifers. Always on your ass about something…" He continued ignoring me for the moment.

"Look, I have to get coffee for the old man, but I found the information you wanted." He paused as he opened the file cover. "You were reassigned yesterday to Camp Eagle in the north, 101st Airborne and you reported yesterday afternoon."

I said "thanks" and got up to leave. He called out for me to wait with a strange look on his face as if he just realized what he had said. That if I reported then why… But apparently the old man's coffee was more important than what ever it was that was wrong about this situation.

He waved me off with a flick of his wrist and threw the file into the pile with the rest on his table.

As I sat watching the "freedom birds" fly overhead I wondered if I would ever get out of this place.

The seriousness of the situation came back when everyone started to line up for chow and I realized that I still did not have a meal ticket. In all the confusion of being inside that building it had slipped my mind to get a new one. I went to the clerk in the mess hall and asked how you acquired a new meal ticket.

"For what?" He asked nonchalantly.

"So I can eat." I said sarcastically.

"Oh hell, you don't need one go ahead."

"But I thought that everyone…'

"That's just another way to screw you around."

He was right. For the rest of the time I was there I continued to use my expired ticket and not one person questioned it. Another strongly enforced army rule, I thought to myself.

Later that night I lay in my bunk, complaining how uncomfortable it was. Little did I know I would not lie in another bed for a long time to come.

One of the new arrivals walked over and started pumping me for information. He gave the appearance of being quite scared. It was amazing, here he was as scared as I was on my first day and here I was laying casually on my bunk watching time fade away, drawing heavily on a cigarette. To him I was an old-timer (the oldest one here at least). At least to him I was an old-timer. I had been here for awhile so that made me a pro on the subject.

I suppose as with anything you eventually get conditioned to the fact you really are in Vietnam. He was pressing me for any information I could give him on what this was all about.

I, in my newly acquired laissez faire attitude, explained that it's no big thing. As long as we were stuck here for a year we might as well make the best of it. He wasn't satisfied with that answer. He kept ranting and raving about the fact people get killed over here. I said very casually "only if you go in that building". Not having the faintest idea what I was talking about, he dismissed me as quickly as he had approached. I continued to draw on my cigarette.

Finally my day arrived and I was reassigned up north, camp Eagle 101st Airborne as a matter of fact. I wondered if I would meet Howard Jordan. He felt like a brother by now.

# "P" TRAINING

**EVERYWHERE I LOOKED, THERE WAS RED CLAY** (dirt). In the hills, on the ground and even all over my damn clothes. The only place I didn't see red clay was in the remaining mud holes that were left from the previous monsoon season. Maybe my luck was changing. I had missed the rain.

We were led to a set of barracks, actually wooden shacks that were set in a circle around a central shack that had running water, obviously used for sanitary purposes. These were also laced with red clay dust.

As we stepped inside it became apparent that these were not going to be much of a paradise, mainly because there weren't any beds, or anything at all for that matter. We were issued sleeping gear, which consisted of a poncho and poncho liner. It looked more like a comforter and a piece of plastic to keep one dry when we slept in the rain. Who the hell would want to sleep in the rain anyway? Boy did I find out.

As I lay there that first night, I started to visualize what "P" training would be like. No one really knew what "P" training meant, but it sounded good anyway. They would teach us to be combat soldiers here. Fight, fight, fight, just like Notre Dame, but they played football not war games, games? What the hell was I thinking about? This wasn't a game anymore. Hadn't they issued us M16s with live ammo? We hadn't even been here a full day. What had that guy said? We might need it. What the hell was happening? It wasn't a joke anymore. I wasn't laughing.

However I was not to continue this line of thinking any further, because the voices next to me were drawing my attention.

"Yeah, I can do it." The first voice said.

"You're crazy." Another voice responded.

"No really, one shot and I can blow the motha fucker to bits."

There went my heart again. Were we being overrun already? Could this really be happening?

"What would that prove?" The other voice answered.

"Don't got to prove nuttin, man, just that I can."

"But what does shootin' the fuck out of a light bulb prove to anybody?"

Huh… What were they talking about, shooting a light bulb? Why would they want to shoot the light out when we were being overrun by VC. That's it. They wanted the light out so they would be in the dark when the enemy attacked. We would wait until they came closer, then we would slink out and catch them off guard. They would not expect us to attack them, when they were attacking us. I grabbed my rifle and ammo.

Nothing was happening. Nobody was preparing. In fact nobody was moving. Something was definitely wrong. Fortunately, before I had a chance to yell charge, I came back to my senses and realized that the two who had been talking were now quite subdued.

"Well, are you gonna?"

"Hell, I'll get it later. I'm tired now."

"What do you mean, tired? Do it."

With that the other fellow handed him the rifle.

"You do it."

The second fellow did not appear to be interested in the subject anymore and let the rifle fall to one side. Their interest returned to rolling another joint.

I reached over and picked up the now forgotten rifle. I proceeded to release the magazine and unchamber the round. I carefully replaced the rifle next to the man. However, I kept the ammo myself. I certainly didn't want anybody "shooting the fuck out of a light bulb" in the middle of the night.

The next morning we were marched to chow. As we were marching, we approached a rather large mud hole still filled with water. Of course everyone marched around it. The sergeant marching us became quite irritated about this, so he started mouthing a barrage of profanities at us. Somewhere in the middle of his sentence structure I caught the phrase, "…you'll sleep in that shit before your tour of duty is over…" There it was again, that reference to sleeping in the mud and rain. Maybe there was some truth in it.

The training consisted mostly of convincing us that all that "horse shit" we learned in the states was worthless. One thing for sure, there weren't any close order drills (marching cadence). We were basically re-instructed in the use of our M16s. The quick kill method.

"Remember men this is…" Holding the gun (God forbid rifle!) as if it were a beautiful woman, caressing it ever so gently. "…this will be your wife, your girlfriend or your mother, whichever you prefer for the next twelve months, so treat it as such.

The next day, feeling a bit weary of all the bull shit, I decided to volunteer for a small duty assignment. Even after hearing

# | 16 | THE PROTECTED WILL NEVER KNOW

all the rumors about volunteering for anything in the army (the call was for assistance in the supply depot), I decided to volunteer anyway.

Several of us, ten, I believe, boarded a truck, content in the fact that we had gotten over for a change. We would not have to join the training today. We would instead have some cushiony job moving supplies. There was, however, the possibility it would be heavy supplies, or some type of tedious job that might require who knows what. Oh well, it still beat training for a day.

As we arrived at our work point, we were met by a short stocky dude packing a forty-five pistol strapped to his side as if he were back in the old west. He was also chomping on what was left of a cheap cigar. He greeted us with what appeared to be a sly grin as if he knew we were about to be "fucked" once again in our army career.

We were herded into two ton and a quarters (army version of the standard pickup truck) for transport to another place. There was something weird about the contents of these trucks. In each truck bed there were six fifty-five gallon drums that were cut in half and filled with JP4 (gasoline). Now what the hell were we going to do with these? The answer would reach us very soon.

We headed down a back road that took us behind the mess hall. So that was it, we were going to refuel the mess tent. Why? Why the hell would the mess tent need fuel? For the generators. Of course, they would need fuel to power the generators so they could cook. We pulled up to a small building in the back. Oh my God. We all recognized it at once.

The short stocky dude was out of the truck and standing beside us.

"Okay, pull the old one out and put a fresh one in, load the old one back up on the truck."

"Do what?" We all seemed to say at once.

"I said, pull out the old shit cans and put a fresh pot in its place. That's simple enough. Ya dumb fucks."

So that was it. We were changing the cans in the outhouses. The initial smell was overwhelming, but gradually we were able to adjust to the shock of it all. In fact we started to joke about the whole thing. Such classic lines as "we got a sick one here" and "there's no constipation in this bunch".

After we had replaced all six we had to drive out to the dump and burn our goodies. We pulled the cans off the truck and carried them down into the dump. When we were back up and clear the short stocky dude fired his forty-five into the can until it ignited, getting some perverse pleasure out of shooting the shit (sorry couldn't resist).

That proved to be an uneventful episode until we realized the flammable qualities contained in our precious cargo. Here we were, smoking all day around this shit (no pun intended) as flammable as it was. Oh well, another close call.

Here I was, twice close to annihilation and had not even seen the enemy. This was shaping up to be a hard war. More on that later…

Sometime during the day of my "shit burning" detail, I was close to headquarters while they were processing some new people. I had the distinction of overhearing the conversation between a new recruit and an old sergeant.

"What did you say son? You're what?"

"I'm asphyxiated by the sight of live ammo."

"Well what da fuck does that mean?"

The sergeant in all his splendor was really "helping" this guy along. The new guy, now a bit frightened, was trying to adequately explain his dilemma to the ever so understanding sergeant.

"You see, when I'm around live ammo, I become very weak and... and... and..."

"And what for Christ sakes?"

"...and I faint..."

"YOU WHAT!"

"I faint. The sight of live ammo causes all my muscles to grow weak, my breathing to become shallow, suddenly everything goes blank and I pass out."

"You gotta be shittin' me. How da fuck did you get to Nam with a problem like that, for Christ sakes? Jeeze..."

"No one would believe me when I told them, so they thought I was lying and sent me here anyway."

The sergeant now shaking his head back and forth turned to the clerk next to him for support.

"Jeeze, what are we gonna do wit this kid now?"

The clerk, who obviously was a company kiss-ass, tuned slowly to the sergeant offering him some thoughts.

"We could assign..." (You mean condemn him I thought) "...to the commo bunker." (The commo bunker was actually a fortress located about twenty feet below ground. It was used for the communications channels within camp Eagle.) "That way he wouldn't have to see any live ammo."

The sergeant appeared at first to be pleased with this idea, but then his face contorted again, as if he had remembered something he forgot to do before reporting for duty.

"But he has to come out sometime."

"So what..." The clerk continued. "...even if he faints here, it won't be as bad as out there."

Again the sergeant seemed content with this line of reasoning. Wasn't this why the kiss-ass clerk was assigned to him in the first place, I thought? I left them as the sergeant was speaking to the new guy.

"Son…"

The rest of my time in "P" training was uneventful, except maybe for the day we got to blast a bunch of fifty-five gallon drums with our M16s as target practice, to learn the quick kill method of combat. One thing for sure, it was better than light bulbs.

# BAN ME THOUT

**THE NEXT STOP ON MY JOURNEY** was back to the south in (or around) Phan Thiet, a nice little resort village right on the ocean. Phan Thiet was down closer to Saigon in the south of South Vietnam. The army had built one of the many base camps right on the white sand beaches.

I was to discover upon my arrival that this would be a short stop, because the unit of the 101st I was to join, the 3/506 (101st Airborne Division, 3rd Battalion, 506th Infantry) was not in Phan Thiet anymore. The unit had been sent back up north to work with an Air Force unit.

There were three of us, who arrived for our assignments and would be here a day or two, mostly to pick up our gear and then actually join the unit, wherever they were.

We had heard back at camp Eagle that the 3/506 was the only unit of the 101st operating in the south in II Corps (two core). Among others, II Corps was the home of the 173rd Airborne Division. As I found out later we were often referred to as: "third of the five oh who?" The other elements of the 101st Airborne

## | 22 |   THE PROTECTED WILL NEVER KNOW

Division were all in the northern region of South Vietnam in I Corps (eye core) and now we find that our unit is in the north.

When we finally arrived at our new base we discovered that we were still very far down in the southern region, but in relation to Phan Thiet we were north again. Basically we were in the central highlands. At this point, it wasn't worth thinking about anymore. It really didn't matter a whole hell of a lot where or how far south/north we were. It was all still Vietnam.

They were right about one thing, the unit was walking guard duty for an Air Force unit around the village of Ban Me Thout (Ban me too it). What the whole assignment amounted to was watching the outskirts of the chain linked fenced camp that had been built for the Air Force unit to do, whatever it was they were doing, in peace.

I couldn't complain though, because the Air Force had real plates to eat off and running water for their showers, but most of all, flush toilets. How long had it been since I had seen one of those? At least a month. Each of us had to pull an eight-hour shift of guard duty after which you were on your own.

Being a new guy, instantly renamed "cherry boy" (for not having been shot at or in a combat situation yet), it quickly became apparent where I stood on the priority ladder. Before I had a chance to set my gear down (which everyone was now calling a "ruck") I was handed an army version of a two-way radio, model number PRC25 otherwise known as a prick twenty-five.

The fellow handing it to me was all smiles and saying how glad he was to see me join the unit. I could instantly tell why he was glad to see me. Even more so when I took the radio from him and felt the added weight. Although it wasn't that heavy by itself, added to the rest of my gear I knew... I knew... I knew we needed another cherry boy soon.

I was to be an RTO (radio telephone operator). It could have been worse, I could have been handed an M60 machine gun which, including the extra ammo you had to carry with it worked out heavier than the radio. You always have to look on the bright side.

As a new arrival, I would not receive a guard duty assignment until the next day, when the rotation had come back around. That enabled me to have the rest of this day plus the next off. Some compensation I thought, for being new.

It was late in the afternoon, so there wasn't that much to do but sit around exchanging hometown and war stories that night. One of the guys decided we should go over to the club the Air Force had set up for its lower ranking personnel. With most of us in agreement we set off to find the place in the totally dark compound.

Fortunately, it wasn't too far, just to the left of their mess hall. When we arrived they were showing a movie in a blacked out building next door, so some of us went there first. After the movie we went into the club for a few drinks. At least they had American beer there.

To cap off the evening's entertainment one of the Air Force guys showed a few very rough stag movies. I hesitantly admit that they were the first I had ever seen (and I'm sure I wasn't alone). Well, things were looking up. This had been the best day I had spent in Nam so far.

The next morning a group of us decided to go over to the "house." The "house" was a converted French Mansion on the main road outside the base. It probably was a beautiful place at one time in its history.

# | 24 |  THE PROTECTED WILL NEVER KNOW

The people running it now had converted the first floor into a bar and dance area. The upstairs, I would find out later, had been converted into a mass of small approximately six by eight-foot rooms. As we entered, we broke into groups with the other two guys I had met in Phan Thiet and myself going off into our own group.

The first problem we faced was money. Jack and I had not drawn any, but Smitty had, so we were negotiating with him about how he could divide it between us. Since he had sixty dollars, it was only reasonable that we should get twenty apiece. Smitty, although very hesitant at first, finally relented and handed it over to us.

Actually it wasn't twenty dollars as we know it, but it was one of those MPC (military payment certificates) pink in color. I had heard a story that when they first introduced MPC over here, a bunch of GIs had distributed great hoards of MONOPOLY® money to the unsuspecting Vietnamese. I was still entertaining that thought and simultaneously twirling the pink twenty when the girls walked up to our table.

Jack wasted no time. As soon as I realized the girls were there, Jack was headed toward the back with one of them. The other two sat down next to Smitty and myself urging us to go upstairs with them, as Jack had done with his girl.

Smitty and I however, wanted to think this over for a minute. We had exchanged glances indicating to each other that we had better just stay down here for a few minutes anyway. After all, Jack had been to this country before, so it was assumed he knew the ropes. We didn't.

A waiter approached and asked what we would like to drink. Smitty and I ordered beers and we ordered something everyone called a Saigon Tea for the girls. There were at least a dozen stories about what the Saigon Tea really was (everything from an aphrodisiac to birth control), but hell it didn't matter, if that is what they wanted so be it.

In a few minutes he was back with our drinks and as I handed him the pink twenty, which was saturated with perspiration, I had wild visions of the waiter going berserk accusing me of giving him MONOPOLY® money therefore cheating him out of his just dues.

The waiter accepted the pink twenty graciously and proceeded to give me my change in Piestras (their equivalent of money). I watched as he counted out the change and noticed right away that he wasn't sure how much to give me. He watched me with each bill he laid down. When he had reached twenty-three dollars (Piestras) I decided, not one to be greedy that was enough. He nodded and left us. Somehow beer at nine o'clock in the morning really didn't do much for me.

The girls kept insisting that we go upstairs now. After all, time was money, especially in this racket. Smitty and I exchanged glances again. It would be all right now. Besides this was an Air Force controlled establishment and they were always pretty thorough. We walked with the girls to the back.

When we arrived around back we found quite a bustling community back there. There were half a dozen girls in various stages of dress as they were changing from one outfit to another. All wore western style clothes, that is, clothes of our western culture rather than clothes from their culture. Right at the base of the stairs was Mama San (she more or less acted as the madam of the place), obviously guarding the passage to guarantee that all fees had been paid prior to ascending that flight of stairs.

We were led, pushed actually, into a small room with a glass-faced door on the front. There was a piece of gutter affixed to the back wall, tilted slightly down to the left acting as drainage. We were to urinate in there. That seemed fair enough considering most of the populace in this country used the front yard to urinate. However this had a two-fold purpose, as we were about to find out.

As we were urinating into the gutter a woman entered and stared at us for a few seconds, spoke something in Vietnamese then left. Smitty and I looked at each other inquisitively for an instant when another girl outside the door spoke.

"Make sure you okay."

"Make sure we're okay?" We both repeated.

"Not sick."

"Oh!"

We decided to leave it at that.

Smitty, the girls and I approached the stairs and Mama San who stopped us abruptly.

"Fifteen dollar two time, eight dollar one."

"One time."

"One time."

At the top of the stairs people were everywhere, going in and out of rooms racing up and down the halls. It looked like a train station at five o'clock.

Smitty and his girl disappeared quickly. We weren't so lucky. We proceeded down the hall, my girl checking rooms to see if they were empty. Each room had a piece of plywood to act as a door. We finally found one.

As soon as we were inside she removed her slacks and panties and started egging me on to join her. Time was money. I undressed and lay on the bed next to her. She had left her blouse on and I was attempting to remove it.

"Hands cold."

"Sorry I'm nervous."

I negotiated the blouse up past her breasts when she stopped me.

"Far enough."

"What...?"

"Mess my hair."

"Mess your hair?"

She smiled.

Her activities during this period convinced me it really didn't matter after all. She had encouraged me downstairs that we would have fun upstairs. That and her instance on calling me cherry boy for an entirely different reason than the guys had stirred up a fire inside me that would show her what this was all about.

Sensing this she threw her arms around me and smiled again. After a few rounds of touchy feely I entered her. Release was instant. She smiled.

She cleaned both of us off. We dressed. She smiled. We left the room and went back downstairs. Now I understood what one time meant.

Smitty was already back down stairs. I joined him while the girls went back into circulation. We smiled at each other, than laughed uncontrollably. Jack still had not returned. Must be "two time". We laughed again.

Since it was only ten o'clock in the morning business was kind of slow for the girls. Apparently all the guys in here were from last night's shift or about to go on their shifts. In any event things slacked off. Our girls came back over by us probably to coax us into another time.

We bought them another tea and ourselves another beer. Smitty and his girl were really going at it, a steady stream of kissing and fondling. I thought for sure they were going to do it again right on the table. The girls left for a minute to freshen up and Smitty remembered I was there.

"I think I'm in love."

"You're what? You can't be serious?"

"I know I'm in love."

"Hey look, she just… a…"

"I don't care, she's terrific."

I had to get him out of here. I thought it was great too, but I wouldn't go that far. The girls came back and Smitty and his girl went back to their kissing and fondling. My girl climbed up on my lap kissing and fondling which I obviously didn't object to. We had to get out of here. Where the hell was Jack?

I looked over a few minutes later and Smitty had removed his girls' blouse and slacks and was trying to maneuver her into a workable position. My girl and I looked at each other. She smiled.

Suddenly we were all aware of Jack's presence. The smile on his face was different, that is, besides the animal lust in his eyes. I finally spotted it. His front tooth was chipped and he was speaking with a lisp.

"Jack what the hell happened?"

"I lost my fuckin' cap somehow."

"But how?"

After I had asked, I realized I really didn't want to know. It was enough that he had left way before us and had returned long after us. I certainly wasn't interested in how he had lost his cap. Just then, Jack's girl walked up, actually staggered up and sat down next to my girl and me. She was still on my lap. Smitty and his girl were all over the table, stopping long enough to look over at us. I looked at my girl. She smiled. Jack's girl smiled.

Smitty's girl smiled. Jack smiled (leered actually). Smitty smiled. I smiled.

We sat around for awhile, bought another round of drinks, and then decided to leave. It was almost time for noon chow.

When we got outside, we walked silently for a few minutes then glanced at each other and we all laughed uncontrollably the rest of the way back to camp. I never did find out how Jack lost his cap, nor whether Smitty was really in love. However, I also decided to let it be. The fun was in not knowing.

Jack and I pulled guard duty together. We swapped the usual hometown stories and general bullshit we could conjure up to pass the time. We spent the better part of the next three days together.

On the afternoon of the third day, Jack was starting to get a little philosophical. Not that those "special" cigarettes had anything to do with it. We had smoked quite a few the previous night, at least Jack did. I had backed off when I started lighting two at a time. Jack had thought that was extremely funny and had entered into an uncontrollable fit of laughter, which we finally had to suppress with a heavy wool blanket.

Later on, Jack had decided that we would be attacked during the night and he wasn't going to spend the night in the tent like a sitting duck. He went out to spend the night in a small little mound of sandbags, which had been placed there in case we were attacked. I found him at three in the morning sleeping in the outhouse. He had decided that if we were attacked he would want to get a healthy shit in first. With a little coaxing I got him back to the sandbags.

"When do you think you'll get it?" Jack asked.

Actually I hadn't thought about it at all. The question had taken me completely off guard. Jack continued.

"Ya know you'll get blown away, at least fucked up over here. It's just a matter of time."

"Jack, I never thought…"

"Com'on don't give me that bullshit. You know you been thinkin' on it since ya got here. So when?"

"Probably February. It's December now, two months."

"Ya really think Feb? Nah, you'll make it to May, June the latest."

"Well that's encouraging. I appreciate your faith in my existing that long. Should I look forward to it? How's your percentage?"

"Aw com'on man. Don't get all fucked up over this. I'm just curious on what ya think. I been eight outta ten right."

"You been what?"

"I been eight outta ten. But look don't worry, not all kills. A lotta wounded. I just got this knack for knowing that's all."

I sat quietly looking at, actually past Jack into the night.

"Look, you and me's been pretty close the last coupla days and I felt I owed it to ya. To let ya know what I think. I didn't mean nuthin' particular by it. Okay kid?"

"Yeah it's alright Jack. I suppose I shoulda given it some thought. I mean people do get killed and wounded over here, but I guess I just figured it wouldn't happen to me. But it's cool. I got no remorse or any of dat shit."

"Hey you're alright kid. Did I ever tell ya about the first time I was here?"

I shook my head no. Among the various subjects we had covered the last three days that was not one of them. Jack continued.

"Well I was working a LRRP (long range reconnaissance patrols — commonly called "lurps") team then. That's where six of us goes out on our own, walkin' patrols and we work different areas, then report back. It's a pretty good deal because we're mostly to ourselves. Know what I mean?"

Jack paused, pulled heavily on the cigarette he was smoking and letting the smoke release before continuing.

"Well anyway, we was pulling this one patrol and we had to climb this big fuckin' hill. I mean this motha was high and damn near straight up. Well anyway, we were working our way to the top moving kinda slow so as not to draw attention from no one and I started to slip. Well with all of this shit I'm packin' I tried to get my balance back, but I can't ya see because the weight on my back shifts pulling me the other way."

"Next thing I know my knee's on fire and I'm laying there like a crippled dog. Well the guys call in a dust off, (medical evacuation helicopter) that's one of them special birds they use for a medivac to get the wounded and dead out. But I wasn't wounded just my knee all fucked up."

"They kept me in the hospital for awhile, but when they realized I had torn some cartilage in it, they had to send me home. When ya got things like that, especially broken bones, they send ya back to the states, 'cause it takes too long to heal. Well that was all right wit me. I didn't want to be here no way."

"Well I get back to the states with all this time left ta do 'cause I only spent a month and a half over here. They got to keep me in. Well, I get assigned to a supply depot in Washington State. Think they'd send me to Oakland, shit no, up in the cold."

"I couldn't handle it. I'm telling ya man, I really tried. I took all the bullshit 'til it was coming out my ears. After being here, stateside duty is really shitty. They fuck ya over every chance they get. Especially when they found out I had just hurt my knee, that I wasn't no big fuckin' hero."

"This one second louie, you know, second lieutenant, really

had it in for me. He kept riding my ass about how I had gotten over by leaving Nam so soon. Hell, the prick hadn't been here himself, but he was riding me. Can ya beat that? He would press me for stories, then brag to the others about how it was over here. Shit, he didn't know his ass from a hole in the ground."

"Well I finally had enough, so one day he starts riding me again, so I busts him up side the head, but the asshole, instead of backing off and throwing my ass in jail hits me back. That's how my tooth gets broke in the first place."

"Well some people sees this and breaks it up right away, but I got him cause there's witnesses both sides. Well, the army don't want this kind of mess, so they offer us new assignments to forget the whole thing and start fresh somewhere else. I don't know where he went, but I had wanted to come back here 'cause they don't have that bullshit like they do in the states. So here I am. How's that grab you kid?"

"Why do you keep calling me kid?"

"Hell I don't mean nuthin' by it kid."

I had learned a little more about Vietnam from Jack's story. He had actually wanted to come back here. Even after what we had just talked about, about when we would "get it". I had to know why.

"Jack, with the outside chance of getting killed here, why didn't you take something in the states or some other country, anything but here?"

"Because everything but here is stateside shit and I couldn't handle that no more. Here or nuthin."

"Yeah, but you could get killed or fucked up or something here. Anywhere else ya got a chance to live at least."

"Believe me kid, after you get used to the freedom ya get here, you just can't adjust to the states. Lemme try to explain it."

"Here they know ya got them, 'cause what's the worst they could do to you? Send you to Nam, right? Well you're here already so that gets rid of that threat and most of all you could get killed or fucked up here and you're thinking that and they know you're thinking that."

"They ain't gonna lay no more pressure on you like they do in the states. No one bugs you here about whether your fuckin' hair is cut or your boots are shined or if you shaved this morning. All they care about is you do what you're told so they don't get blown away or fucked up because you fucked up."

"One other thing man, you're carrying a loaded rifle. Ain't nobody gonna fuck ya over with that in your hand. Hey you been takin' too much shit as it is and some lifer-dog-pig-fucker who's decided to make a career outta being fucked over all the time by staying in the Goddamn army his whole life ain't gonna give you shit 'cause he knows you'll turn that rifle on him and blow him away and he knows that. That's why they don't fuck with ya over here. Can you understand what I'm telling ya kid?"

"Yeah Jack. Yeah, I can handle that. I know what you're saying now."

"Yeah it's simple kid. Just remember the simple premise. You don't fuck with no man carrying a loaded rifle. They won't and you don't either."

"Say Jack…"

"What kid?"

"Thanks."

"Yeah. It's cool. Want a hit?"

"Yeah, I think I need one."

"Go ahead. It helps. Hang in there, kid. You'll make it. I don't see you buying it."

"Buying it?"

"Yeah. Dying. You'll just be fucked up a little."

"I'll tell you what. Light another one. We'll get fucked up together."

"Sounds good to me."

Jack had put Vietnam, and the whole war for that matter, in a different perspective for me. I was beginning to understand things a lot better. I had already lost one phase of my cherry boy status and I was slowly losing the other.

Here we were, the United States Army, the finest if you listened to what the people in it tell you. The scum of the earth if you listened to what the other branches of the service tell you. Fighting a war in a Southeast Asian country called Vietnam. What we were supposed to be doing was helping the southern element defend itself against the northern element, which amounted basically to a civil war between the two elements. However, in the middle of all this we were fighting the Viet Cong, a guerrilla faction that was fighting everybody. Lay in the bullshit we had to endure over here and it became one of those who were the enemy here situations.

Will the real enemy please stand up?

Although none of that was supposed to matter. There was a principal involved. This was the democracy of the free world against the communism of the imprisoned world and this was to be a proving ground. Or some shit like that. Maybe there was a principal involved, I just didn't comprehend any of it. What did I know? I was just a poor kid from the south side of Chicago who had been drafted and reported here because I had received orders to do so.

To further complicate matters, we were receiving reports from the states about how the majority of the populace was against our involvement here, period, no matter what the reasons.

It seemed that you were damned if you do and damned if you don't. What the hell could you do? I understood more and more of what Jack was saying. This was the best place for those of us in the army.

Will the real enemy please stand up?

With all of this going on in our heads, we also had to deal with some representative of the army, be it an officer, a sergeant, or anybody with a higher rank than you, ready to harass you in some way. Who needed it? I was beginning to understand how easy it would be to just turn your rifle on someone hassling you and pull the trigger.

I was beginning to feel a strong friendship building between Jack and I.

"Jack."

"Yeah?"

"Have anymore?"

"Yeah, ya want another?"

"Yeah, I need it."

"Hey kid, is what I said getting you down?"

"No Jack. It's just that I'm realizing what you said."

"Hey, I didn't mean to scare you or nuthin."

"No Jack. Really it's okay. I guess I just been takin' things too light."

"Hey kid forget it. Ain't nuthin' you and I can do to change things. It was like that before us and it will be like that after. Here have another hit."

"Thanks. When did you realize all of that heavy shit?"

"I never really did kid. I just been hashin' some things over on my own and started puttin' two and two together. It all seemed to fit, so I let it be that way. What do you really think, kid?"

"Up to now, I didn't think anything really. I just kinda went along for the ride. But I gotta admit, what you said made a lotta sense."

"It's the joint, kid. It helps you relax, put your mind at ease. Just let the sensation happen. You'll feel better."

"What about our guard duty?"

"We were relieved an hour ago."

"We were?"

"Yeah, but it's cool. Those guys are smokers too."

"Jack, think we ought to head back?"

"Why we can be a head here."

With that we again broke into uncontrollable laughter. It would be all right after all.

I am not quite sure when we went back to the tent, but after a couple more joints I decided to spend the night outside the tent as Jack had the previous night. It was more peaceful anyway.

We spent two more days in Ban Me Thout, a week in all. I had learned a lot from this small village or at least from being assigned around it. A lot of growing up had happened in the week I was here. A certain coming of age had overtaken me and I felt a certain affection for this place, my first real assignment in Vietnam.

It was with deep personal regret that I watched on the news in 1975, the capturing of Ban Me Thout by the North Vietnamese Army as they swept through the South toward the end of the war.

No other place left as strong an impact on me as this place did. I had learned a lot here. Not just of the war, but of life and death as well.

Farewell Ban Me Thout.

She smiled.

# MY FIRST FIRE BASE

THE UNIT WAS NEXT ASSIGNED TO WORK **OPCON** (operations control) with the 173rd Airborne Brigade somewhere around Ah Khe further up in the central highlands near Pleiku. Although the 101st Airborne Division was stationed up north in I Corps (eye core), our element, the 3rd of the 506th was further down south in II Corps (two core) working with the 173rd Airborne Brigade. This time it wouldn't be so easy. We were assigned to start building a firebase on top of a mountain peak.

It was the highest of several mountain peaks grouped in this area. My God, the army did something right. An artillery unit that would set up their big guns on the very top would also support us. We would construct bunkers made out of sand bags around the lower perimeter. It didn't seem too bad of a job.

The area seemed like it had been inhabited before. The jungle type climate that was prevalent throughout the country had a way of quickly overgrowing everything, but this area had definite signs of previous habitation. So much so that the jungle had not overgrown it. Had they tried this before, I wondered?

# | 38 |  THE PROTECTED WILL NEVER KNOW

I let it pass for awhile thinking that there were probably good reasons for being here and having been here before.

As we started digging in we found various pieces of the previous inhabitants, including an ammo box containing two-year old Playboys. We also found the remains of some sandbags along with c-ration cans and other odds and ends usually associated with grunts in the boonies. Well, okay, so we had been here before, so what?

As the days pressed on, we got to wondering more and more about the units that were here before us. No one really knew why or what had happened which only caused our curiosity to rise even more.

One day Sky Power (Chaplain) came out to visit us. I guess it was a Sunday. He was a Protestant Chaplain but we all went to his service, even us Catholics. We thought out here it really didn't matter who spread God's word. They all looked the same to us, standing in the mud and dressed in a uniform. Maybe they should practice that way in the states. It certainly would alleviate some of the competition between religions.

At any rate after the service we pressed him for details about what this place had been before. He hemmed and hawed, not really saying anything. We pressed harder. He had to know something. His avoidance of the subject only made it worse. Finally it got to him. He decided to tell us what he knew.

"Now listen boys, I don't want you to get the wrong ideas about what I'm going to tell you."

"No father we won't," someone said.

"Father? He's Protestant," someone else chimed in.

"So what, that's close enough."

"Now boys we are all God's children." The Chaplain interrupted.

"Well, us children want to know what the hell happened here anyway."

"Com'on watch the language. He's holy."

"I'm sorry."

"You're sorry, alright."

"Hey…"

"Alright boys enough. I'm sure there wasn't anything meant by it."

"So what happened, Father?"

The Chaplain looked us over as he gathered his thoughts.

"Well every year there is a strong buildup of enemy forces in this area and with these hills it's hard to keep a good watch on their movement."

"Oh, so we just stay here for awhile until the buildup moves on. Just to sort of break up their action. Is that right, father?" someone interjected.

"Well not quite boys. You see each year this action starts to build around Christmas time. I guess they believe since Christmas is a holiday we would slack off a little and they would be able to catch us off guard."

"That ain't too hard to figure, but with us on the hill they know it won't be too easy cause we'd be watchin' what they was doin'."

"Well, yes, something like that," the Chaplain replied.

Everyone seemed to be content with that idea. It appeared logical enough. Everyone, that is, except Jack. He needed to know more.

"Fodder, you're not telling us something."

"Yes son, I've told you what I know."

"Fodder, com'on we can take it, we been around."

# | 40 |  THE PROTECTED WILL NEVER KNOW

Jack was definitely in command of the situation. He had the Chaplin perspiring. He must be holding out.

"I don't know what you mean, son."

"Fodder, the army don't just build a base once a year then tear it down and rebuild it for no reason. What else happened here?"

"Listen boys, I don't want you to think it can happen all the time."

"Fodder, what happened?"

"Well, last year on Christmas eve the base was annihilated."

"Shit… some Christmas…," someone yelled out.

"Hey…"

"Yeah… sorry, Father."

After everyone digested his words they seemed to accept it could happen. This was Vietnam. Jack still pressed.

"What else Fodder?"

"I told you last year…"

"Yeah Fodder, what else?"

Jack had actually put the Chaplain into a cold sweat. In the middle of one hundred degree heat the poor Chaplain was shaking all over. Even I could feel the cold steel of Jack's eyes piercing through the Chaplain. I had to feel for him. This would probably be harder than his day before God. Jack pressed harder.

"Fodder, you're stalling. We want the rest."

"Well, as I said, last year on Christmas the base was annihilated. Well, the fact of the matter is, it has happened every year for the last three years."

"Holy shit," someone yelled.

"What day is it," another guy asked.

"Sunday," the Chaplain replied.

"Yeah I know that. What's the date?"

Everyone stood silent as they tried to calculate the actual date. One more or less lost track over here. The days, weeks, months all appeared the same. The Chaplain saved everyone the trouble.

"December 21, 1969."

"That means…"

"Four fuckin' days to obliteration."

"Hey, the man's still here. Have some respect."

"Yeah, sorry Father. It's just…"

"I know son. I can understand your feelings."

Jack still wasn't satisfied. He had to know more.

"Fodder, how do they do it?"

"I don't know about other years, but last year they mounted big guns and mortars on each of the other hills and shelled this one for twenty-four hours. They started Christmas Eve and continued until Christmas night."

"But didn't the guys here get support?"

"Yes, but for each gun they knocked off another was put in its place. There was just too many to fight off."

"Merry fucking Christmas…"

"That's a lot to look forward to. Thanks, Father."

"Boys, I tried to tell you…"

"Fodder, it's okay, we appreciate your telling us."

Jack was in command again. He led the Chaplain back to the top of the hill to wait for his helicopter back to base camp.

When Jack returned we were all huddled discussing our strategy to prevent it from happening to us. Jack seemed to be a little less confident now. Even he had been affected by the Chaplain's story. However, not one to lose his cool, he immediately took charge again.

"Look guys, they don't fuck up all the time. I'm sure they been thinking about it already. They probably got something figgered out on how they gonna take care a it."

I had to agree with him. I mean three years in a row? Somebody had to realize something by now. Oh well, it was four days away. Back to the business at hand. We had a firebase to build.

The next day we all watched as fighter jets from the Air Force bombed each hill repeatedly. They concentrated on the tops first, then hit each of the four sides about one hundred yards down each side. They repeated this performance every morning and every afternoon until Christmas Eve.

After they had finished their runs, which were actually called sorties, we would send a team in to check for any remains of enemy positions. This gave those who were airborne qualified a chance to practice their trade.

Due to the bombings the ground was chewed up pretty bad and they wouldn't land helicopters, so teams would go over and repel (jump) from the helicopters via a rope into the bombed out hill tops. It did the job.

With all the attention we were giving to each hill it appeared highly unlikely that the enemy would be able to build up anything this year. We were hoping anyway.

On one of the days we spotted two-suspected VC on the hill across from us. They had binoculars and appeared to be watching and noting everything we had and everything we were doing. We had a bead on them and were ready to blow them away, but had not received the order to fire yet. The CO (commanding officer) had called Battalion Headquarters for permission to fire, but they had not responded yet.

I asked why we needed permission to fire. Weren't they obviously the enemy? Wasn't this a war where survival meant everything at all costs? I was informed to the contrary. Yes, this was a war, but there were also innocent civilians involved in this too. We had to be careful who we shot. They had developed such a thing as a "no kill" and a "free kill" zone. This was a "no kill" zone.

As we sat there waiting, the CO decided to send out a team to capture them. They couldn't complain about that, he yelled. However the team would have to work their way down our hill, and then work their way back up the other hill. It would take quite a bit of time. The enemy would not wait for us.

One of the guys in our platoon was a crack shot. In fact he carried an M14 instead of an M16. The fourteen was a more accurate weapon for precise shots. The sixteen was for the quick kill. He kept a continuous bead on the two just waiting for the order to shoot.

The CO decided he would send two teams out, one down the left, the other down the right. I was in the team to the right.

We were working our way down the hill in a much quicker pace than if we were out on patrol. We were trying to make time. It took us a half hour to get down. It would take us a lot longer to get up. The enemy would not wait forever.

What kind of war was this, I wondered? Why would you need permission to kill the enemy? It was obvious what they were doing. We couldn't afford to let them bring that information back to anybody. It could mean our lives. We pushed harder. We had to get up there and catch them. Word came over the radio.

"Com'on back. They left."

I removed the handset from the radio and replied.

"They what? We're almost there."

## THE PROTECTED WILL NEVER KNOW

"Forget it. They're gone. CO says to com'on back."

I motioned to stop and told the sergeant the news. He stopped the team, had us take a break and get ready to head back. I could hear the traffic on the horn. They were informing the other team of the news.

We had failed. They would bring that information back to their units. We would be the fourth year in a row. We rested. There would be plenty of time to go back up.

When we got back to the top I looked at the rest of the guys. They seemed as dejected as I did. They all thought the same thing. We would be the forth. Word had finally come through. In answer to our question as to whether we could fire. The answer had been no.

Will the real enemy please stand up?

After the Christmas Eve morning bombing and after the last of the checking, mine implants had been laid into the wooded area just below the bombed area. We all sat around waiting for dark to come over the camp.

It was one of the many aspects of this country. It would be light for twelve hours then dark for twelve hours. Usually it was from seven to seven. It would be dark soon.

We had heard it was harder to get support in the dark. That seemed logical enough. But at this point logic was not prevailing. We would be attacked soon, or so the story went, and we would need that support.

As night approached everyone began to tense up. Nerves were on edge and what was usually good-natured fun became irritating child's play.

Once night fell we all prepared for the inevitable to happen. When it did not happen by midnight a few started to dismiss the Chaplain's story as a ploy to get us back into his religious world.

Still others said it had to be the precautions they had taken prior to Christmas Eve.

Hadn't we bombed those hills every day so even if the enemy would reinforce during the night, the sortie the next morning would eliminate that? Whatever the reasons we were all glad it didn't happen. We had broken the streak. We had driven them off this time.

The rest of the night passed a little easier. Some of the guys had even been able to get some sleep. We had survived another day. I caught up with Jack during the night.

"Say Jack, I was beginning to wonder about your percentages. I was hoping I wouldn't be your eight out of ten."

"You and me both, kid."

After my hour of guard duty I went back inside the bunker and tried to get some sleep. They army would not let us have Christmas off. At least not out in the boonies. I was restless. Sleep would not come. I decided to go out back and have a smoke. Jack was out there as well.

"What' the matter kid? Can't sleep either?"

"No Jack I can't. Thought I'd have a cigarette."

"You want a hit of this?"

"No, thanks. Don't feel like gettin' high. Rather stay down right now."

"Ah kid listen, that's the problem. You're keeping your head filled with all this bullshit flying around and the possibility of our being hit. You need a little something to help ya relax, put your mind at ease for awhile. You still got a lotta time left here. Plenty a time to think all that shit. Here go ahead, have one."

Jack was right again. I needed to forget things, for awhile at least. I did have a lot of time left. I shouldn't start worrying now. Worrying was usually reserved for the short timers.

## THE PROTECTED WILL NEVER KNOW

You started counting the days left in your one-year tour of duty the moment you set foot in country. Then you set a benchmark for yourself, when you break three hundred or one hundred and eighty "on the down side" of the count. Once you break one hundred you are a "two digit midget". At thirty days left you start angling to get out of the boonies. Don't want to press your luck. At a week left you usually start processing out.

Some guys said they were "short" as early at the one eighty break, but most considered themselves short when they broke ninety. Anyone under thirty was "short". Short timers were the ones to get nervous. They surely didn't want anything happening to them that close to leaving.

I definitely wasn't one of those with only a month and a half in country so far. I hadn't even broke three hundred yet. The best I could say was I had forty down. A count backwards theory until I at least broke three hundred. I still had ten and a half months to go. That is unless I believed Jack.

"That's right I shouldn't worry 'till May… or June."

"What… oh yeah, that's when you'll get it. See, don't worry until then. Here now, have a hit."

I took the joint from Jack and drew in deeply, keeping the smoke down as long as I could before releasing. I took another hit. I finally lit my own. Three joints later I was feeling better already.

"Say Jack, how come these seem much better this time?"

"Better than what kid?"

"You know. Better than the first time I smoked?"

"That's 'cause the first time ya just wanted to be one of the guys so ya smoked for the sake of smoking."

"You could tell it was my first time?"

"Yeah, but it didn't bodder me none. If you wanted to smoke you'd smoke, if you didn't ya wouldn't. It didn't matter one way or the other."

"Weren't you worried that I might fink or something?"

"Nah. It's that loaded gun theory again. You wouldn't knowin' I could blow you away any time."

"You would have done that?"

"Of course not, but you don't know that. I'm as new to you as you are to me. We gotta trust each other 'cause we're in the same boat. See what I mean kid? It all comes down to ya gotta trust somebody over here. Everyone else is against you."

"But you didn't let on that I was new at it. You let me stumble through without saying nothing."

"Look kid, everything is new to somebody sometime. It's the only way ya learn how's things is. Can you dig it?"

"Yeah, I can dig it. But these do taste or should I say feel better this time."

"You got it right kid they do feel better this time 'cause you smoked for the pleasure of it instead of smoking to be one of the guys like you did the first time. You knew the feeling it produces so this time you took it for that 'cause you were down. Believe me man, you were way down when you came out here."

"I suppose you're right, Jack."

"Of course I'm right. Have I steered you wrong yet, kid?"

"No, not yet, but I'm still uptight about my getting it in May. Should I count on it? I've been pacing my mind to believe I still have ten left. How good was your percentage?"

"Eight outta ten. Don't forget that two."

We both laughed until it hurt. We smoked some more before turning in. It must have worked. I vaguely remember going back in.

## | 48 | THE PROTECTED WILL NEVER KNOW

Christmas morning was announced by the screaming of an off-key chicken. I'm sure the state I was in had a little to do with it, but I actually felt like I was on a farm and I was thinking (dreaming) we had to hire a voice coach for that chicken. I could not take another day of his off-key croaking. Do chickens croak? Well, whatever they do it had to be taken care of.

The heavy artillery people on top of our hill had taken up the ritual of harassing us with an off-key rendition of the opening lines of the *Chickenman* episodes that played on the radio. The *Chickenman* was a bumbling type of character who became a hero through his mistakes each time.

The emblem of the $101^{st}$ Airborne Division was a white eagle emblazoned on a black patch ("Screaming Eagles"). Others often referred to the patch as the puking buzzard. However, this artillery unit decided a chicken was more appropriate, hence the relation to *Chickenman*.

Now each morning we awoke to the cry of "Bawk, bawk, bawk, baaaawwwwkkkk, it's *Chicken Man* he's everywhere, he's everywhere", the opening line of each episode. Of course it was sung in a high pitched, off key voice usually with several others joining in for the finale.

We had grown accustomed to it, but this morning it was worse than ever. What with it being Christmas, we begged mercy of not hearing the finale just this once. They sympathized and the whole gang up there joined in for the finale, which damn near shattered eardrums.

If there were any enemy in the area they would surely leave now. Who would want to fight that? So be it. It was Christmas!

Quite a few of us had received packages from home that contained various sundries; such as canned ravioli or spaghetti and just about anything else they put in cans. We had all decided to save everything until Christmas day and cook up a big stew. What the hell it had to beat c-rations.

Our meals to this point since leaving Ban Me Thout consisted of packaged can goods the army had labeled as c-rations. These usually consisted of a variety of items such as ham, mostly fat, some kind of pork and spaghetti, which I swear looked just like maggots crawling over brains.

There were also cans of fruit, crackers, candy and jam. In each unit there was a package of necessities, a wad of toilet paper, a book of matches, a pack of four cigarettes, a plastic spoon, coffee and sugar. In all sincerity, it wasn't too bad an assortment.

We were also supplied with a sundry box that contained various kinds of cigarettes, some candy and a variety of other goodies. Resupply was usually every three days so it wasn't that bad. One other item we received, were heat tabs. They were some sort of burning agent that could be used to heat your food.

In order to cook we would take one of the used cans punch holes in it with a church key (bottle/can opener) place a piece of heat tab inside, ignite it and it would produce a low flame to set the canned meat over for heating.

Of course we experimented with different ways of heating and cooking, but at the time, that was the best way to get in a quick hot meal. One of the guys had used a piece of C4 once with disastrous results.

C4 is a high explosive when used in conjunction with a detonator. It is actually a white, clay like substance that can be molded into any form, which is what makes it so useful. However when applied to a flame, even a lit cigarette, it will burn at a very fast and very high rate.

## | 50 | THE PROTECTED WILL NEVER KNOW

Well, this one guy had wanted to save the heat tabs he had left and used C4 for cooking. It probably would have worked, but he used too big of a piece and when he lit it under a can of ham it literally melted the can and fried the ham to a cinder. We used C4 sparingly after that episode.

For this Christmas feast we were about to cook we decided to use our steel pot (the army's name for our helmet). It fit over the plastic piece we would adjust to fit our head. It also had a cloth covering colored in green patches to act as camouflage.

As most did I wore mind backwards with the flat side to the front instead of the back. I had also printed the word Chi-Town (Chicago) on it to act as my identification.

Due to the fact it was made out of steel we had decided to use C4 as our heating agent. The heat tabs would not be adequate to keep this big pot brewing. Of course we would use quite a few sticks, but we considered it the army's contribution to our stew. It was Christmas.

Rumor had started to circulate that we would be allowed to go down to the stream at the bottom of the hill to take a bath. That was really encouraging. It had been awhile and even though we all smelled the same, we smelled. We would also be issued fresh clothes. Maybe this would be a merry Christmas after all.

To complete the package we would be able to eat a hot turkey dinner that was being flown out especially to us. Our little group would miss that though because our turn to eat and take our bath came at the same time. We couldn't do both. We decided to take the bath. Besides, we had our stew brewing.

The stew proved to be a huge success. Afterwards we sat around smoking and trading stories. The fear of being annihilated had vanished and was replaced with the routine cutting up we all did.

In a few days we packed up and moved out. We had won this year. The group taking over next year would be able to rest a little easier knowing the annual Christmas firebase was on a winning streak.

It seemed a little ridiculous to tear down the bunkers we had just built a few days ago, but we couldn't leave anything for the enemy to get a hold of. Down they came.

We watched as the bigger Chinook helicopters flew in and carried out the big guns, dangling from each as they left. We would be the last to leave. We had to make sure the guns got out safely.

It was sad, in a way, we had worked hard to build this base, now we had to make sure everything was destroyed. Oh well, another episode finished without incident. I might make ten more after all.

# MONSOON SEASON

**I suppose it had to happen sometime.** We were assigned this time to an area that was being saturated by monsoon rains.

Monsoon rain was one of those things you had to experience to fully appreciate its devastating blow to the morale. The rain would come down in ten-minute (or so) spurts, and then disappear for ten minutes (or so) before it returned. I don't think anyone really timed it, but ten seemed like a good average.

The rain itself was more like a fine drizzle. It was just enough to keep you wet all of the time. We would build huge bonfires to try and dry off. However, as soon as you were dry, the ten minutes would be up and the rain would start again. You could actually see the rain approaching. It was kind of freaky in a way, to stand there watching as the rain began approaching, knowing there was nothing else to do but stand there and get wet (or wetter).

The hilltop we had landed on this time instantly became a quagmire. The mud went from patches ankle deep to damn near knee deep in just a couple of days. The daily battle was one of trying not to disappear into a hole never to be found again.

## THE PROTECTED WILL NEVER KNOW

One consolation, it was very soft to sleep in. "You will sleep in that shit before your tour of duty is over." That's what the sergeant had said in "P" training. Well, I had finally been exposed to all those warnings about sleeping in the mud and rain. Monsoon season did not let up at night.

It was one thing to have to stand out in the rain steadily getting wet, but to have to lie down and try to sleep under the same circumstances is a completely different experience.

A couple of the guys had elected to carry wool blankets in addition to their poncho and poncho liners. Can you believe the added weight of a soaking wet wool blanket? It damn near did a person in to try and carry that thing. All were abandoned immediately. Besides how much heat could a cold wet blanket offer?

It is amazing how much one takes their body for granted. Our bodies are equipped with a built in mechanism called body heat. This reacts with any covering we would supply to keep you warm and safe from external conditions. Under most circumstances that is a perfect combination.

However, it was raining now. Add in the breeze that was blowing to keep the wet coverings cool and you have an external condition that becomes a bit harder to beat.

Well, believe it or not, the army actually created a variation on this natural phenomenon, the poncho and poncho liner. The poncho was a plastic covering with a hood in the middle to wear when one was forced to stand in the rain. Being plastic, water rolled off and it would keep whatever was underneath dry.

By lying down and placing the poncho on top of you, you kept what would be new rain from reaching you. There was another feature to the plastic covering. It would also keep the heat in.

What in fact happened was, your body heat would heat up the wet poncho liner actually creating steam as the water became heated, thereby keeping you quite warm while you were under it's protection. Due to the fact the steam didn't have a chance to escape, being under the poncho it became absorbed back into the poncho liner. This gave you the feeling of sleeping in a warm bath all night. I had to admit, wet or not, it was definitely warm.

Occasionally your lying in the mud would cause you to sink, winding up in a puddle, but as long as you stayed covered the water was also heated. It was really amazing. Sleeping in the rain and mud wasn't so bad after all.

There was one major drawback, however. In the morning you had to get up from under your protection. The cool crisp morning air, although it was not really cold, raced across your nice warm body with a fury that could only be described as that of stepping out of a hot shower into a snowstorm. Without that protective covering your body (and clothes) still quite wet became re-exposed to that external element again. You immediately raced over to the bonfire for heat.

Standing by the bonfire would also heat the water off you, but only one side at a time. Facing the fire the front would become dry while your back was being saturated from the next ten minutes of rain or vice versa. You couldn't win.

Another of the many drawbacks to monsoon rains was support. Or should I say the lack of it. With the rains moving in ten-minute intervals there was zero visibility. Helicopters would not fly if they couldn't see.

It had been decided that under these circumstances we could not accomplish anything nor if the enemy did attack would we be able to do much about it. Really what defense did we have all huddled around a big bonfire, with our rifles tucked away so they wouldn't turn to rust? If the enemy wanted to they could just walk in and capture us, anything to get out of the rain.

The decision had been made to pull out of this area. Too bad we didn't have a way out. Without the helicopters flying there was no way to take us out. We certainly weren't going to walk down. You just can't go down a hill in the mud especially with about one hundred pounds of gear on one's back. It would be disastrous. If one guy slipped we would all go. Well, we would just have to wait it out was the word from higher up.

The first couple of days it was still good-natured complaining, but by the third day everyone had had enough. By now everything was wet. Ever try taking a crap, then using wet toilet paper to wipe yourself? Doesn't work. Take my word for it. You couldn't light wet cigarettes especially with wet matches. Even the joints had to be saved for future use. Hell you couldn't even get high. Not to mention your skin.

Since the first time you got wet, which was within the first twenty minutes after arriving, your skin remained wet. Every part of your skin became pruney, the condition that happens when you stay in the bathtub too long. Well, that had happened to all of us by now.

Coupled with that fact, were patches of jungle rot starting to spread all over your body, specifically in the areas where something rubbed against your skin. Usually around the area of your watch as the band slid up and down, elbows, knees, crotch and especially your feet, that were inside wet socks and wet boots. I wasn't sure what it was until Jack noticed it on me.

"Hey man, you're rotting."

"I'm what?"

"Rotting. Ya got jungle rot on your wrists."

I had noticed huge patches of raw skin peeling off each wrist and had been shifting my watch back and forth between each wrist, as the irritation would subside on one.

"Is it bad?"

"Nah kid, as soon as ya dry up it'll go away."

"What the fuck is jungle rot anyway?"

"It's just from being wet alla time."

"How do you keep from getting it?"

"Stay outta da rain."

"Com'on Jack I'm serious. I got this stuff all over."

"Yeah, so do I kid. Look it's nuthin'. You get it from something rubbing your wet skin. The wet skin isn't as strong so it is more susceptible to being worn away. The first couple a layers ya don't notice, but when it gets to showing red ya worry a little, knowin' ya don't have too many left to rub away. But look when ya take the pressure off it starts to heal itself. It leaves a scar but that clears up after awhile. Just cool it kid. Ya need a smoke. For that matter so do I."

Because of the rain, none of us had been able to smoke anything, let alone a joint. Fortunately, I hadn't become that dependent on it yet, but even I was feeling the urge. Smoking a joint had helped me over the other crises I had been through so far and I was sure it would help now. Talking to Jack would help as well.

"Say Jack, why do they call it jungle rot?"

"How the fuck should I know? I'm not sure they do. That's what we call it and it's good enough for me.

"Christ, sorry I asked."

"Forget it kid. Guess I'm getting a little sick of this shit. I wish we'd get outta this fuckin rain."

"Yeah, me too. They said maybe tomorrow they'd get a bird in."

"In this shit? Look out there, it hasn't let up for three days. What makes them think it will tomorrow?"

"Hell, I don't know. That's what I heard."

It was three more days before they could get us off the hill. We spent New Years out here. During guard duty on New Years Eve I heard someone whisper "Happy New Year" over the radio it was followed moments later by a not so quiet "Fuck You". So we made it to 1970.

We were brought back in out of the boonies for a day to dry off and eat a hot meal. We were not able to cook in the rain either.

We spent the next day just scraping mud off and out of everything. It had been a mess getting out. We tried to carry everything we could, but I'm sure there were things that had been buried in the mud, which had been overlooked and just left there. It was nowhere near the clean up job we had done when we left the firebase. Oh well, if the enemy wanted what we left, they would have to dig it out of the mud.

## MOVING OUT

**ONE OF THE PROBLEMS OF NOT BEING WITH OUR OWN DIVISION** was not having a home base to go back to once in awhile. We would always be out in the boonies. When we did go in it was always for short periods, usually to re-supply, change clothes, clean up our rifles and leave again.

Most units would be able to get back in after a few days or a week or so, but we didn't really have a place to go back to, just the small setup in the rear for support. So our home was really in the boonies. Most of us didn't know better as it was that way from the beginning.

It was probably better this way not having anyone hassle us. When you were in the rear, you had to shave every day, get your haircut and all the other army crap you usually had to deal with. Not to mention dealing with the REMFs (Rear Echelon Mother Fuckers), from officers to enlisted personnel whose job it was to provide support and services to those of us humping the boonies. They had their world and we had ours.

In our case, because it was a short "stand down", we would only be there for thirty-six hours so no one usually hassled us.

They just wanted to get rid of us right away. The next day we were off again.

Whenever we went to or left a place we did it by means of a CA (combat assault). We would climb into Huey Helicopters, five in each. These were small units with two pilots and two door gunners, one on each side. We would pile in, one inside, two on each side hanging off the edge. The trick was to get in and out as fast as possible.

When we would go into a new area, the helicopter would hover about two feet above the ground staying at full idle until all five of us were off. They would usually land four at a time if it were possible, bringing our whole platoon down at the same time. Otherwise they would land in whatever formation they could, from one to two at a time. What you had to watch out for were ground forces waiting for you when you landed and got off the bird (helicopter).

Once we were on the ground we were on our own. That is of course, if there wasn't any trouble. As soon as the helicopters had cleared out we re-grouped and prepared to "hump the boonies" (walked through the thick jungle with fully loaded rucks) to a new spot. Hovering helicopters made a good marking for a mortar barrage. We would move out as fast as we could to a spot we could reasonably set up in and secure.

If it were close to dark, we would find a place where we could spend the night. If it were still light enough, we would continue moving.

Basically what we were doing was running patrols, better known as "search and destroy" missions. In simple terms we would "search" for enemy and if we found something we would "destroy" it, or have it destroyed through artillery or air support. The idea was to hit the enemy before they hit us. Problem was, whenever we found something, or captured a hill or area, we always gave it back and moved on to the next hill or area. So we were always "on the move".

We would be given one "click" to cover. Another one of those terms that seemed to pop up all the time. A "click" represented one square on the map. Each square was used to mark off 1,000 meters of land. I had heard the term "click" came from the sight of a sniper's scope, one click per thousand meters. We would have to cover this square.

There were four companies in our battalion. Each company had three platoons. Each platoon had twenty to twenty-five men. We were operating in platoon strength. Our company had been assigned a particular square or click to cover and each platoon would in turn cover their assigned squares or clicks.

As a platoon we would set up in a particular area of the square then send out teams of four to six men to patrol around seeing if they could find any trace of enemy infiltration in the area. They would go out so far sweep around and return. Once they returned another team would go out and do the same. When we had covered a specified area we would pack up and hump the platoon to a different location and start over.

We would maintain constant radio contact with the other platoons in the area to avoid a possible run in with them. You certainly didn't want to fire on or be fired at by your own men.

In addition to possible enemy contact in the area we also had to deal with the elements, both natural and man-made.

The man-made were much easier. You could remove the threat by either destroying the obstacle or leaving the area. These obstacles consisted mostly of pungi stakes, pungi pits and various booby traps.

## THE PROTECTED WILL NEVER KNOW

The pungi stakes were pieces of bamboo sharpened at one end and sunk into the ground at the other. Sometimes they were dipped in stuff to heighten their effectiveness, but usually they were just positioned to cause the greatest damage to anyone that ran into them. They were placed such that by mixing with the jungle growth and angled so as to instantly penetrate your lower leg upon contact, rarely did they ever kill anyone, but they certainly put one out of action for awhile.

The pungi pits could kill, because they were actually deep holes completely filled with pungi stakes. If you were to fall into the pit, depending on how he fell the stakes would go right through him.

The booby traps came in all shapes and sizes from simple grenades with a trip wire to large unexploded bombs wired with additional charges to create double explosions. These were usually placed along paths or areas that were used for travel so someone walking by would snag the trip wire and detonate the explosive. They were also found attached to anything a grunt might pick up or touch. Everything was booby trapped, from documents, to bodies, to hooch's (grass shacks built by the villagers), to abandoned equipment (theirs, ours or ours left by the ARVNs [Army of the Republic of Vietnam]).

The natural elements were much harder to cope with simply because they would not go away and were everywhere you went. The list was endless, but some of the most persistent, were heat and humidity, jungle and insects.

The heat and jungle humidity were constant. It was always over one hundred degrees and damn near that close in humidity. As most did, I stopped wearing underwear the first week I was in the boonies. That left me with a fatigue shirt and pants, socks and boots. We were re-supplied every three days and hopefully we would receive a fresh set of fatigues. After three days of perspiring the fatigues were pretty gross. In many instances rashes would appear around the heavy perspiration areas. If we happened to be near a stream we would wash out the set we had, but that was not often.

The jungle growth caused many problems in our movement. We would have to chop our way through some rough terrain (most of us carried machetes, two foot long knives). Of course it was advisable not to follow any formed paths or well-worn trails. Occasionally we would run into "hold-me-back" vines, they had a way of wrapping around our gear and at times would actually stop you from moving until he was cut free. This was especially true for guys like me that carried a radio.

The most annoying of the elements, were the insects. They were in abundance everywhere. The two most prevalent were flies and mosquitoes.

The flies would bother us during the day. I suppose because of the dead and decaying bodies around they had ample chances to breed. I had observed maggots totally devour a dead wild boar. I did not believe they took exception to a human body.

On those occasions we did find a dead gook body (name we gave to the enemy), the maggots would have made camp. Between the heat, the flies and the other insects attacking it, I was told a body would "go to bones" in six days. If the body was badly damaged from artillery or a direct hit of some sort it could go faster. The flies were the first and the maggots were the cleaners.

It was interesting to note that back in the world had a fly landed in my coffee I would have poured out the coffee and got a fresh cup, but in Vietnam, out in the boonies with water as precious as it was, I simply fished out the flies (and whatever else landed in my cup) and proceeded to drink the coffee. Amazing how one's standards change.

The nights belonged to the mosquitoes. Almost as soon as it became dark they would start their rampage. We used an insect repellent, which hopefully would keep the mosquitoes off of us. It was quite, but not totally, effective.

We slept completely clothed, including boots. We would draw our poncho liners over our heads to cover our faces. Even with that we could still hear them buzzing around. At times, it sounded like dive bombers making a run on a target.

With all these mosquitoes around you had a great chance of catching malaria. We had to take two different kinds of pills to protect us. Some of the guys had opted to not take the pills, hoping to catch malaria. It was a good way to get out of the boonies for a month.

Some of the other less bothersome insects included centipedes, termites, black ants, bees, butterflies and roaches. There were scorpions and tarantulas, but other than a few minor chance meetings they did not pose much of a problem. We did have a guy who was stung by a scorpion but he was back in the boonies three days later. Banana spiders were killer, but we only ran into those if we were near a banana grove.

We did occasionally run into snakes, but a blast from an M16 or a quick swish of a machete solved that problem. We also had parasites in the form of leeches and ticks.

We set up in a damp area that had been saturated from the monsoons. It was near the end, and although it wasn't raining it was still very wet. The deeper we crept into the jungle the wetter it was. We had just begun to set up when the first leech was spotted. It was ignored until several others were spotted. If it hadn't been so close to dark we would have moved to another spot, but we would have to stick it out here.

Everyone was becoming "buggy" from the fear of contacting leeches. In the morning we all did a strip act to check each other out. Some guys did in fact have leeches on them. We were able to get them off with lit cigarettes. Once a leech attaches itself to your skin it would have to be coaxed into moving. The heat of a lit cigarette was a great coaxer. I was fortunate; the leeches had not bothered me. However, ticks were another story.

As is common with not bathing regularly you develop a habit of continually scratching, especially in the groin area. My groin scratching had become rather painful. I first suspected I had developed a rash and decided to have a look. It did not appear to be a rash. It looked clear enough.

Upon closer examination I found it. A HUGE (huge in my eyes) black ball was growing behind my testicle. I immediately screamed for the medic. He informed me that it was a tick and from the looks of it, it had been there awhile. I casually (as casually as any man can standing there holding his testicles which now numbered three could be) mentioned that I would like it removed.

The medic asked if I had a cigarette. Instantly realizing what he wanted it for, I backed off and shook my head no. He went for his scalpel.

The thought of his cutting me in that delicate area caused me to back away again. I was now standing there with my pants down around my ankles and my hands cupping my groin. The medic convinced me I had three choices. The lit cigarette, the scalpel or the tick staying. The scalpel won.

Two guys held my arms; to make sure I would not attack the medic at an inopportune time. I watched as he made an incision on either side of the black ball then on the top and bottom. That was as far as I could watch.

In an instant the medic was standing holding the black ball in his fingertips. He had removed it head and all. Fortunately, the head had not traveled too far in. I breathed a sigh of relief. As a parting gesture the medic asked if I wanted a band-aid. I politely declined.

The most devastating insects we ran into were red ants. They would build large complexes that, when we discovered them we would blow up or otherwise destroy. They had a tiny bite, but when multiplied by the hundreds that would attack at once it became effective. Majority of the time we would spot their complexes and avoid them. Sometimes they would set up in trees. We would not notice and inevitably someone would bump the hive knocking them down. Of course the guy behind was deluged with them. We would immediately stop and help the guy get them off.

One time we had caught a scorpion, a tarantula and a centipede. We had them in a little hole trapped. They just kept crawling around each other. One guy decided to add a little excitement to the party. He had trapped a bunch of red ants somehow and threw them into the hole. The ants immediately attacked the other three bugs, devouring them in seconds. We plugged the hole with C4 explosive and ignited it. That was the best way to handle red ants.

Of all the other insects and/or animals we encountered, none bothered us as much as these. The red ants were definitely the fiercest element.

There was always something to look for in each click (square). Although we always had the heat, we did not always have the other elements. Some areas would be free of everything but flies and mosquitoes.

When the entire click (square) was checked out we would move on to another. Sometimes we would hump to the next and sometimes the helicopters would come in and fly us to the next.

As a matter of record we were making a lot of combat assaults (CAs), but in reality we were just moving from place to place. I suppose though, that if one of these areas were hot (that is had enemy present) it would have been a true CA.

Occasionally we had to land in some pretty rough terrain and although the helicopters did the best they could we would sometimes have up to a six-foot drop to the ground. Normally that might not be so bad, but when you had about one hundred pounds of gear on your back it becomes a bit more involved. A couple of guys came up with sprains and one guy broke a leg.

So far, that had been the only reason to call in a "dust off". These were medical evacuation helicopters (medivacs) that came in to get the wounded out. In my opinion they were the best support unit in Vietnam. Those pilots could park a helicopter on a dime to get the wounded out. They had pure guts when it came to removing injured men.

On another occasion we landed in a rice paddy. After jumping from the helicopter we landed knee deep in whatever it is in rice paddies. Again, because of all that gear, it was hard getting back out.

We were all beginning to feel like a band of gypsies. We kept moving from place to place. We finally wound up in a hilly region that promised to keep us in one spot for awhile.

It was a little harder to work the hills. Up and down. Up one side, down the other. When we were finished we would pack up, go over the top and work the other side.

It was during one of these relocating to the other side jaunts that it happened. We were moving along a very rocky region and it was very hard keeping a sure footing. A couple of guys had already slipped with no serious injuries.

I was moving slowly along trying to keep my ill-fitting ruck from sliding all over the place. Its weight kept shifting precariously from side to side disrupting my normal balance ability. I don't actually remember slipping. The next thing I realized I was tumbling forward, actually free falling down the hill.

One of the guys further down had stopped my forward motion and another broke my ruck free of my shoulders. That was a nice feature of the ruck, it had a quick release strap. As I lay there I felt a tremendous pain building in my left shoulder.

The company called in a dust off (medical evacuation helicopter) to take me out. We were on the side of a hill and the dust off couldn't land so they decided to use a jungle penetrater to pull me out.

The jungle penetrater was a seat suspended from a long cable they could drop from the helicopter to pull me up. It was somewhat difficult getting into that chair, but after a few minutes, I was adequately strapped in and was being hoisted up.

That wasn't exactly the way I had planned to leave, but I really didn't have much choice. I kept thinking about Jack's percentages. It was only January.

Once I arrived at the field station, I was immediately rushed into a room where a nurse was waiting to prep me for a doctor's arrival. An x-ray unit was wheeled in to take pictures of my now throbbing shoulder. The doctor arrived within seconds viewing the x-rays as he entered.

"What happened son?"

My speech was somewhat labored as my shoulder continued to throb.

"I fell off a cliff."

"With full gear?"

"Yes."

"I believe you lucked out son. Now hold on, this is going to hurt a little."

With that he began to press down on my throbbing shoulder until something snapped. It was amazing that what was building to be great pain was now slowly subsiding.

"It will be tender for a couple of days, but no serious damage. You can go back to your unit tomorrow. Spend the night here, relax and take a couple of these, it will help you sleep. You need that more than anything else right now."

"What happened, doc? What was it?"

"You just jacked up your shoulder a little and I pushed it back down."

"That's it?"

"That's it. You want all the medical jargon that goes with it?"

"No thanks, doc, I'll take your word for it."

There's Jack's percentages again. I had pulled through once more. Wow, I would sleep in a bed tonight. That wasn't too bad. I'd rejoin my unit tomorrow and everything would be back to normal.

I flew back to my unit in the morning. Of course they were surprised to see me pull up. We had a small setup in the rear specifically for the paper work and our re-supply. It was only two tents set off to the side in a compound called LZ (Landing Zone) English.

After I explained why I was there and what had happened, they shrugged and let it pass as if it had happened everyday. I was instructed that as long as I was there I would get a haircut, change into a better fitting uniform (the one I had on had been tapered to remove the looseness that accompanies army uniforms, in addition I cut the sleeves on my shirt to allow my arms to breathe) and shine up my boots. Jesus, now I knew what Jack was talking about. However, I was without a rifle. It had been confiscated back at the hospital. They had something against loaded rifles inside a protected compound.

I had only been here a couple of hours and I was ready to go back out to the boonies. I would have to wait a couple of days though until the next re-supply. They would not fly me out special.

I spent that time hiding and building a new ruck. As was the procedure, once a person was medivaced out, his gear was dismantled and distributed among the remaining guys. Whatever they didn't want was either sent back in or field destroyed.

I rounded up ten more canteens, that's what I had been humping, a new canteen cup, another 20 magazines of M16 ammo and three day's worth of food (c-rations). I would get my radio back when I rejoined my company. I was somehow sure no one would want to keep it.

This time, I kept rearranging my gear until it felt right on my back. I even borrowed an extra radio to affix to test for balance. When I was finally satisfied with the fit, I tucked it away and waited to head back out to the boonies.

They would not issue another M16 until I was ready to board the helicopter. That was okay, I didn't need it yet, or so I thought.

I spent the next day just trying to kill time and stay out of trouble. I only ran into one hassle. I wore a crucifix around my neck that I refused to remove for any reason. I had worn it since I was a senior in high school. I would not remove it now.

There was a major that I understood was the "XO" (executive officer) for battalion headquarters that had decided he was fed up with all the paraphernalia the guys were wearing around their necks, especially peace symbols, which I was also wearing.

When he spotted me he immediately ordered me to remove that "shit" from around my neck. I obligingly removed the peace symbol, hell I could put it back on later, but staunchly refused to remove my crucifix. He became enraged at this and started shouting things like I could be court-martialed for my insubordination and what have you. That didn't matter, I was not removing that crucifix and I would like to see him try to court-martial me. It became a shouting match with the major firing out threats and me daring him to do something about it. Finally a captain who was probably assigned specifically to take care of things like this came over and broke it up.

Where was my rifle? Goddamn REMFs.

Will the real enemy please stand up?

One of the guys from my unit came over and pulled me out of the crowd and walked me over to the other side of the building. He explained that it was in my best interest to forget about it. The fact that I was wearing a blank uniform (we wore uniforms from the laundry pool, none of them had any markings) and that I would be leaving the next day convinced me it wasn't worth it to make an issue out of it.

A little while later, a special order came down forbidding the wearing of anything around the neck, religious or otherwise. The order was posted and copies made to be sent to the grunts in the boonies.

I clutched the crucifix hanging from my neck and swore I'd go to jail before I'd take it off. I wanted to see them try to put me in jail for wearing a symbol of my religion. Not that I was any sort of religious fanatic or anything like that, but now it was a matter of principle.

I would not have to worry long. The next morning when that major came out and tried to start his jeep, he discovered it would not start. When he lifted the hood he found to his horror that two one-oh-five millimeter shells had been wired to the ignition during the night. The shells, which in comparison to the engine looked like he had two engines, had been miswired either by accident or on purpose just to pose a threat and not to actually detonate.

We imagined that had the shells detonated, it would have taken with it most of the Quonset hut it was parked next to. Whatever the case the plan worked. The major applied for an emergency leave back to the world and left that afternoon. I guess the major pissed off the wrong person or persons.

The order was taken down and the copies were ripped up. I put my peace symbol back on.

I rejoined my company that afternoon. It was good to be back. I immediately changed into better fitting clothes from the re-supply sacks. I got my radio back, as I suspected, fit it into place and was ready to go.

We would move into position for the night and get as close to Hill 474 as we could. We would be working 474 tomorrow and we wanted to cover as much ground as possible today.

Along the way we had to cross a stream. We always hated doing that late in the day because we would not have enough time to dry off before nightfall. We looked up and down to see if we could find a shallow spot to cross. Fortunately we did. Everyone made it with just getting ankles wet, everyone that is except Benny. Somehow he hit a drop off and sunk to his chest. We got him out right away, but he was still soaked.

After we secured the perimeter, we set up in a grassy knoll for the night. This was accomplished by placing trip flares, which are small explosives with a wire attached that when detonated would give off a bright light to illuminate anything standing there. We also set out claymore mines, which are plastic encased bee bees that when detonated by hand explode, sending thousands of bee bees shooting into the area that had been entered and was now illuminated.

It was quite effective for striking the first blow. We sat around waiting for dark to come. As usual the procedure was to wait until about nine o'clock to start pulling guard shifts. Each guy would pull an hour shift all through the night. It was a standard routine. We did it every night.

This night we were getting ready to turn in when the trip flare popped. The claymore was blown immediately. For good measure a grenade was thrown into the area of light. We had a good view of the area. We watched. Then we saw the shape.

The animal raised its head with a mighty roar. It was bleeding profusely from its left side. We had wounded a tiger. A tiger had wandered into the trip flare and was stalking in the now unprotected area. We decided to toss out a couple of more grenades to either kill it or scare it off.

With this accomplished we heard it roar in the distance. Hopefully it would go away. Sleep would not come easy.

We had to get some sleep, tomorrow would be a rough day. We would have to climb 474. We would really be moving out tomorrow. For some of us it would be the last.

# CONTACT

THE FOLLOWING MORNING, AFTER FINISHING MORNING CHOW, we headed toward Hill 474. It was another one of those rocky hills. Everywhere we climbed we hit massive rocks.

We were beginning to realize that these huge rock formations could act as hiding places for the enemy. The whole hill seemed to be a maze of huge rock formations. There were large chunks of rock torn out where artillery shells had made point blank hits, without doing much damage to the structure as a whole. A feeling of apprehension was beginning to build in everyone as we moved more and more cautiously up the peak.

Hill 474 and its surrounding hills were located northwest of Bong Son in Binh Dinh Province. It was just a few miles northwest of LZ English, a large base camp for the 173rd Airborne Brigade. As we were working opcon to them they would provide support if we needed it.

We had been assigned to work this area mainly because of its proximity to LZ English. There had been reports of enemy buildup in this area to mount a front against the base. We had been working this area for a week without incident so far. We hoped our luck would hold out.

The day before, a couple of the guys started wondering where these reports originated. We had not even seen a trace of enemy movement let alone any enemy. Hill 474 would be the last sector we would have to check. If this proved clean, we would move into another set of hills behind these.

We filtered into an area of rocks that had formed a sort of ledge cut into the side of the hill about halfway down 474. Another platoon was working the top half of 474 with our other platoon working the far side. We definitely had this hill covered.

We were amazed at the natural shelter or hideout this structure was providing. It was set deep in spots and protruded in others. The whole thing was more or less protected by elephant grass and various other jungle growth. It was really a nice little place, ideal to set up for noon chow break.

We set up the posting of guards to keep watch while some of us ate, then rotated so each person had a turn to eat. We were again amazed at the ability you would have in defending this natural fortress. Each side had a spot one could affix himself into to pull an effective guard post.

The hardest spot to defend against would be the top where we would have to fight from a downhill position, but by actually staying inside the structure we could hold out for days. It would be virtually impossible to penetrate. With that in mind we rested a little easier.

There was still a job to be done. We had to check out this area, too. One of the guys had the reputation of being our resident tunnel rat. It became his job to slither down into the cracks in the rocks that could act as caves.

The NVA (North Vietnamese Army) and especially the VC (Viet Cong) had a reputation of hiding in caves. They would move at night while we rested and hide out in the daytime when we moved.

If we could not find them maybe we could at least find a trace of their existence in the area. Down the tunnel rat went.

The rat had checked out several passageways without success, before he found one that opened into a larger cave right below the main floor. Before proceeding, the rat grabbed a forty-five pistol and a flashlight. This one did not have natural light filtering through as the rest did.

He was gone for what seemed like an eternity. It was already past noon. We should be moving on. As nice as this place was we still had a job to do. We had been unsuccessful so far. It appeared this would be a failure too. There were many areas still to be checked.

By this time I had been assigned the platoon sergeant's RTO, which meant I constantly had the handset portion of the radio next to my ear. The senior or lieutenant's RTO liked to be free of monitoring constantly and with my being next in line, I was usually stuck with the job. There was constant communication between the elements in the area and as expected traffic on the horn (as we liked to refer to the radio) was quite heavy.

The CO was pushing for us to abandon this search and move on. I relayed the message to the lieutenant and his reply was, "In a few minutes." I relayed that back to the CO's RTO.

I heard a cry that the rat was out. I looked over in the general vicinity of where he had gone in, but he had emerged in an entirely different spot, close to where I was standing as a matter of fact.

## | 78 | THE PROTECTED WILL NEVER KNOW

The rat was carrying an enemy ruck. He had found something and the ruck was some of the gear that was down there. The place had to be a gook headquarters (gook was the label we placed on the enemy, period). He went back to the original opening with the sergeant and lieutenant joining him.

The rest of us were perched on the rocks watching to see what else he would bring up. He had dropped the enemy ruck by me for the sergeant to review. I had placed it next to our own gear for a later look, as I was also anxious to see what else this cave would produce. We all waited patiently for his decent.

In the next instant the whole place erupted into one big explosion of gunfire. We had made contact. I grabbed the horn and shouted into it that we were in contact. Instantly everyone reacted. We had been expecting something to happen and this would be handled with minimal effort. Suddenly a voice cried out.

"Guam's hit… medic… medic…"

"How bad?" Someone yelled back.

"Don't know… he's bleeding all over the fucking place."

"Alright stay cool. Guard that side we'll bring him out over there."

The sarge was in charge directing movement. The lieutenant yelled over to me.

"Kid, call a dust off."

"I'm on it, sir."

The firing intensified. The area I was around was being hit with sporadic ricocheting gunfire so I decided to join the rest up front. The gunfire was coming from uphill. We had been caught. Only one man hit. Not bad. We could hold.

The radio was buzzing. They were bringing our other platoon from the top to work their way down and force the enemy toward us. Gunships (helicopters with a heavy arsenal of guns and rockets mounted) were being flown out to assist. The battalion commander was in the area. He would make the pickup of the wounded man with his helicopter. Everything would work out.

The CO kept asking for the lieutenant, but he was directing the positioning of the wounded man for the dust off. After he got things coordinated the lieutenant joined me. His RTO had gone with the sergeant to meet the helicopter. Smitty was handling the radio on the left front (he was carrying the reserve radio) and I on the right.

Smitty came on the horn and said he was taking fire over there too. I brought the lieutenant up to date. He contacted the CO and gave me the horn back. I continued to monitor the horn as I listened to the lieutenant.

"The CO says the third platoon is making their way down to us on the left and the second platoon is coming around the right side. After we get Guam out we should work our way up the hill forcing them into one of us."

"Work our way up? Is he crazy? Charging uphill? Who does he think we are?"

"Soldiers kid, here to do our job."

"Christ sir, be serious. We can barely walk up this fuckin' hill let alone charge. Why don't we just wait for the other platoons to force them down to us?"

"The CO says there's probably just a couple up there harassing us. See how the sergeant's doing."

"Just a couple, then who's doing all that shooting? Two guys can't put out all that firepower."

The rat, who was also our M14 carrying sniper, was putting on a show of his own. I had been breaking M60 machine gun belts of ammo into single rounds for him until the lieutenant arrived. I had seen him hit three myself. He was agreeing with me.

"The kid's right sir. There is a hell of a lot up there and they're entrenched. The last gunship pass didn't even faze them. I don't think we should leave. There's another climbing over that rock…" We heard the explosion of the M14 fire. "Shit I missed…" The rat fired again. "Got him… got the mother fucker."

The lieutenant was transfixed on the rat's shooting. He definitely saw the situation in a different light now. He took the horn from me.

We were using letters for our call signs instead of numbers because of all the other units in the area, especially elements of the 173rd. These would be changed every month to continue the deception to the enemy. Occasionally, however the enemy would find out anyway and harass us over the airwaves.

The month of January we had been assigned various names. Our platoon was assigned xray which was also the lieutenants call sign, delta, the sergeants call sign, alpha the RTO's call sign and mike the CO's call sign. When I answered the horn for the lieutenant I would answer as xray alpha indicating that I was the lieutenant and not the sergeant otherwise I would answer as delta alpha. Likewise, it followed suit throughout the company.

When we contacted each other we would talk RTO to RTO. Once we raised each other we would pass the horn on to whoever wanted the call placed just as an executive would do through his secretary. It was all routine. We also knew each other by our assignments.

Since the professor, the lieutenant's RTO, had gone down to the LZ (Landing Zone, a place to land the dust off helicopter) with the sergeant we had switched call signs. I was now acting as the lieutenants RTO. Although that may not seem very important it was necessary for the other units to realize that when I answered I was actually with the lieutenant.

When the going got tough we would inevitably drop all the preliminary call signs and just answer, so by associating my voice with the lieutenant they would be aware later who was who.

"Mike this is xray over." The lieutenant was trying to reach the CO.

"Mike this is xray, come in, over."

"Yes mike... well, I looked the situation over and... a... it doesn't look too good for us to go up the hill at this time, over."

"I understand mike, but we're taking a lotta shit down here, over."

"Right... yes... yes... I'll get right back... yes... out."

"Here kid, call everybody, get a status report. I'll be back in a minute."

The lieutenant handed me the horn back and left. I grabbed the horn from him and watched the rat pick off another one on a rock he was climbing over. I asked him how it looked out there. He shook his head.

"We're not going out there."

I called the second platoon to see how they were doing advancing toward us. They had two hit and stopped while they waited for a dust off too. I called the third platoon. They had not made it around the ridge. They had been pinned by heavy fire before they even got started. No one hit yet, but the rocks prevented them from trying to move any further at this time. Christ, I thought, what was happening?

I called mike's RTO to see if I could find out about our dust off. He informed me that the battalion commander's helicopter had been "shot up pretty bad" and it had to limp back to LZ English. But they had called for another... so to hang in there.

Hang in? Hell, a helicopter had been shot up. My heart was starting to race again. Maybe this was it. Jack, fuck you and your percentages.

My mind was jumping. Who should I call next? The sergeant, he should be told about the dust off.

"Delta, this is xray alpha, over."

"Delta, this is xray alpha, come in, over."

"Delta alpha, this is xray alpha over."

No one was answering. Did they leave their radio up here? Where was the professor? He had gone down to the LZ with the sergeant. Why didn't he answer? Where was Smitty? He was supposed to be with the left front.

"Delta alpha, this is xray alpha, come in prof, answer me, over."

"This is the professor, over."

"Hey prof, what the hell is happening down there?"

"Down where?"

"The LZ damn it. What are you guys doing?"

"I don't know. Smitty went down to the LZ. I'm halfway down. We had to leave. It got too hot for us. What's happening over there?"

"Smitty...? What do you mean you had to leave? Where's sarge?"

"The sarge is down there too. We're getting hit pretty heavy over here. Where's the lieutenant?"

"The lieutenant went over by you. Can you raise Smitty?"

"Negative."

"Smitty is probably getting static from these rocks. Can any of you guys get down there?"

"We're all going down, there's a lot of shit flying here. I'll get back to ya kid."

"Roger."

All that shit we had been taught about radio language had fallen by the wayside now. This was for real. All hell was breaking loose and we were right in the middle of it. I looked for the lieutenant. He was coming up the side. I tried to raise Smitty again. No luck.

"Give me an update, kid."

"Sir, the second platoon's pinned with two wounded. The third's pinned, no wounded yet. The colonel's bird's been shot up. They're sending another bird out. The left front's been driven out. I can't get anybody on the LZ. In short, we're fucked."

"Christ, we better get outta here too. Get the rat let's go down there."

I tapped the rat on the shoulder and he motioned for me to wait a minute. I watched as he hit another one, on the run this time. He was yelling back at me.

"There are a lot of those fuckers out there. They still want us to charge? Dumb fucks."

"No."

"What?"

"I said no."

"What? I can't hear you?"

"For Christ sakes, NO."

I was screaming at him. He still wasn't responding so I grabbed him and pointed to the LZ. He nodded in response, scooped up what little of his ammo was left and started to follow the lieutenant and me down to the LZ. We picked up what was left of our right front and started inching down. The gunfire was still quite heavy with bullets ricocheting all around the rocks. The horn started to buzz. It was sarge.

"Hey kid, where's that dust off we were promised, we really need it now. We're getting our asses shot off down here."

"How bad is it, sarge?"

"Bad, real bad. We're taking fire from 360. They got us pinned bad…"

"Break, break, this is dust off…"

"Gotcha dust off. We're on a ledge. Can ya make it?"

"Roger xray pop smoke so I can identify, over."

"Roger, smoke going out. Pop smoke sarge, bird's coming in."

"Got it buddy, smoke going out…"

"Roger xray, got your smoke, coming in…"

There was a burst of gunfire.

"Xray, this is dust off one niner. We've taken heavy fire, bird's crippled pretty badly, afraid we'll have to pull out. We'll get another on the way to ya, over…"

"Roger one niner, hope ya make it back."

"Roger, good luck xray, out."

"Sir, the bird got hit, they're going back. Sarge…"

"I heard kid."

The radio fell silent now. We had to get out of here. We managed to get around the right side of the rocks. A couple of our guys were positioned here. As soon as I got in there with them my radio went dead.

I had hit a pocket in the rocks that prevented my radio from functioning. I grabbed the lieutenant and told him of my problem. He pointed for us to keep moving toward the LZ. He was going to leave a few guys here to protect our rear as we advanced.

We approached an opening about 20 feet wide. I was carrying the radio on my back for mobility, but I was not about to cross that space with that thing advertising who and what I was.

I motioned for the lieutenant to go first then I would toss the radio to him on the other side and follow. His trip across was greeted with a burst of heavy fire. I decided to wait a few more minutes before I crossed. After all, what was the rush? My decision to cross was halted by the radio's buzzing.

"Xray, this is dust off…"

"Gotcha dust off, this is xray, come on it."

"Sarge, pop smoke. Sarge pop smoke."

Oh hell I was close enough. I yelled at the top of my lungs. "POP SMOKE!"

It worked. I heard the smoke grenade pop and saw the smoke start to drift up into the air.

"Roger xray, we got your smoke, we're coming in…"

Another burst of gunfire.

"Xray we're hit…we're faltering… have to make emergency landing in the valley. Will send another later."

They were gone as fast as they had arrived. No one could get to us.

"Xray, this is mike over."

"Mike, this is xray alpha. Xray's not here."

"Roger xray alpha can you get to him?"

"Mike, I'm trying."

"Okay kid hang in there. We'll getcha out."

I had to cross now. I had to get to the lieutenant. I took the radio off my back and flung it across the opening. As soon as it hit, I was behind it and diving for cover on the other side.

I rolled into a couple of guys as I hit and lay there for a second to make sure I could move. The first guy I saw was the professor. His pants were soaked in blood. I stared at him in horror. He assured me he was okay and pointed to what was left of his radio.

## THE PROTECTED WILL NEVER KNOW

The radio had been shot up pretty bad. The professor explained that when he crossed, his radio hadn't gone far enough and before he had a chance to retrieve it they had opened up on it tearing it to pieces. I grabbed my radio and hugged it. That sergeant in "P" training was wrong. THIS was my girl not that silly rifle.

I looked for the lieutenant, but he had continued on down. I had to join him. Someone yelled up for more smoke grenades. I grabbed a bunch and started down to the LZ. I headed for a little path that led down from the rocks to the LZ. I had gone about ten feet when I froze.

Right in the middle of the path was a body, a GI body, one of us. I was panic stricken, I could not move. I just kept staring at the body. I had not seen anybody dead before, gook or GI. The sight of this lifeless body lying there right in the middle of everything just tore me right out of the action happening around me. I was brought back to reality by the screams of one of the guys to the right of the path.

"Get down. Get off the fuckin' path. Get outta there."

"Huh… what…?"

"Get down… that's where everyone's been getting it right on the fuckin' path… that fucker up there is good…"

"But what about…"

"Forget it. He's dead."

"But…"

"Get off the fuckin' path…"

I dove to the left. I landed behind a little wall made of rocks. I lay there just staring at the body. The left arm had been wrapped around his head and I could not make out who he was. I kept wondering. Smitty? The medic? Who?

I reached for my radio, which had dropped in the path when I jumped. I tossed the smoke grenades over to the guys on the right to pass down to the LZ. I had to get back to the business at hand. I looked at my watch. It was after three. We had been at this for over two hours. Where was that help? I had to find the lieutenant.

"Where's the lieutenant?" I yelled.

"He went down to the LZ," someone answered.

"Do they have a radio down there?"

"Nah it got busted up."

I tried yelling. "Sir... Sir... Sir can you hear me?"

"What kid?" Came his response from down below.

"Mike's been trying to get a hold of you. What should I tell him?"

"You handle it kid. I can't get out of here. Where's the dust off? We need it badly."

"Right sir. I'll see what I can do."

"Mike this is xray over."

"Go ahead xray."

"Mike we need that dust off badly. The lieutenant's pinned, asked me to relay."

"Roger kid. This is mike. Look, what's your situation down there?"

"Pretty bad mike. They forced us out of the rocks. We're all down on the LZ and we're pinned good. What's the word on help? We're takin' fire from all sides."

"Roger kid. Would artillery help? If I put some big guns on those rocks above you, could you make a run for it?"

"Negative mike. If you did that you'd start a landslide right on top of us, besides we got wounded. We wouldn't be able to move too good with them."

"Roger kid. I get the picture. Hang in there. I'll see what I can do. Oh kid stay cool, you're doing fine."

"I'll try mike."

I lay back for a minute. I had to take a breather and collect my thoughts.

Who is that? God I wish I knew. God I'm glad I don't.

The face had turned yellowish and flies were starting to light upon it. The maggots didn't even wait for the body to be cold before they started their assault. I threw a rock at the swarm. It worked, it chased the flies away for a moment, but they would be back.

I began to wonder what else I would find down here. No, I shouldn't think that. He was the only one. He and Guam that is, but Guam was only wounded. He was down there somewhere, in pain, but still alive. My thoughts were interrupted by voices on the radio.

"Kid come in. You there kid? Kid this is mike come in."

"Yes mike."

"You had me worried there kid. You okay?"

"Yes mike. Just had to take a breather, sorry."

"Okay kid, hang in there. Just wanted to let you know there's some more gunships coming to cover for the dust off. What's your current status?"

"We are moving down the hill toward the LZ. We were chased out of the rocks and we are trying to regroup down here. The lieutenant went down there, but got pinned and I can't get to him but we're in voice contact. Anything you want me to tell him?"

"Not yet kid. The dust off should be in the area shortly. When the dust off arrives get the wounded on and get the hell outta there. We are going to pull back and blast the fuck outta this hill. Hang in. It'll be over soon."

"Roger mike."

The radio fell silent. It would just be a little bit more and we would be out of here.

Who is that lying there?

I told the guys on the right to spread the word that the dust off would be here soon.

It was almost four. We would have to move soon. It would be dark in three hours. How long do these things last? I lay back again. It had been quiet for a few minutes. There was movement to my side.

I flipped over, released the safety on my M16, gripped the trigger and waited. The movement continued. The guys on the right side of the path also noticed. They prepared and waited. The guy who yelled at me to get off the path took charge.

"Who's there? If ya don't answer in two seconds we're gonna blast the fuck outta ya."

"GI...hold it... don't shoot... GI... GI..."

"Damn man answer somebody. You coulda got your head blown off. We don't know who the fuck is out there."

It was the guys the lieutenant had left behind to cover us. They joined me on the left of the path with one keeping a watch on the area he had just come from. They had used the break in the action to make their getaway.

I decided as long as there was a break, I would try to get a status. With the firing stopped it was somewhat quiet now and I could get an update word of mouth.

The professor and his group were still up the path by a group of rocks. The rat was also with them. The sarge was still down on the LZ with the group he took down. The lieutenant was also down there. There were three guys on the right of the path and I had two with me.

The position I occupied was halfway between the professor and the LZ. It appeared the entire platoon was in one of these three places. I decided it was best for me to stay where I was. At least I had voice contact with everyone.

I wondered what happened to Smitty's radio. Smitty? Was that him?

The gunships were really putting it on the hill now. They had arrived and after we had marked our position with smoke they started firing all around us trying to keep the enemy pinned long enough for the dust off to come in and get the wounded out.

The gunships themselves were actually the same type of helicopter that was used for combat assaults and dust offs. These had automatic, high-powered machine guns mounted on each side. They also had rocket launchers affixed. They could really put out the firepower.

The gunships were usually quite effective against the enemy, but this hill had natural protection built into it. All of these rock formations made it difficult for the gunships to really do any damage. This time however they would continue to fire until the dust off could affect its rendezvous with us.

As I lay there watching, I heard the dust off calling. It would be coming in. I yelled for someone to pop smoke and looked in the sky for the helicopter.

I heard a lot of yelling and scuffling from the area of the LZ. I tried to see what was happening, but could not see around the little bend the path took toward the LZ. I tried to raise Smitty on the radio, but there was no answer. I tried yelling to find out what was happening, but in all the confusion, no one down there heard me. Finally, word came up.

The LZ was on fire. The smoke grenade had not popped, but had actually exploded and ignited the elephant grass covering the LZ. I immediately contacted the dust off and informed him of the situation.

The dust off pilot replied that he would fly back down into the valley and would attempt another pass in a few minutes, but we had better hurry. They couldn't wait forever. The gunships would remove their cover. I yelled back to spread the word to the LZ, to do something quick or we'd lose the bird.

Word came that the fire had been contained and they had moved further down and over. They would pop another smoke. I contacted the dust off and he prepared for another pass. He would be coming up and in fast. We should be ready. I sent word down.

Suddenly the dust off appeared coming up out of the valley. At first all I saw was the bottom of the helicopter. But then the pilot leveled off and was making a slow steady approach to the LZ in the area of the smoke.

I was watching with fascination when a burst of gunfire slammed into the plexi-glass windshield actually rocking the helicopter with its impact. The helicopter stalled and rocked from side to side while the pilot tried to regain control. I actually thought it was going to crash into us. I could hear the engines whining as the pilot fought for control. In an instant the helicopter was gone.

The pilot had pulled the helicopter out of the spin and was heading back into the valley. I tried to contact it, but there was no answer. One of the gunships answered my screaming pleas. He informed me that the dust off helicopter had landed safely in the valley and after a few adjustments it would attempt to fly back to English.

The gunship pilot also informed me that he and his crew were running low on fuel and ammo. He would also be heading back to English shortly, but he would stay on the scene until the last possible minute. Further, he was not sure they would be able to get another crew out here before dark.

Dark... I had forgotten about that. It would be dark soon. It was almost five thirty. What would we do when night fell? Obviously we would not be able to get as good support as we had been getting. I knew I had to get hold of Mike.

"Mike, this is xray over."

"Xray, this is mike alpha, mike says wait one, over."

"Xray, this is mike over."

"Roger mike, what do we do now?"

"Is that you kid?"

"Yes."

"Kid ya been doing a helluva job down there. Just hold up a little longer. We're gonna getcha out. Can you get a hold of the lieutenant yet?"

"He can hear me, but I don't think we can get together yet. He had to move the LZ when the old one caught fire."

"Okay kid, listen. Find out how bad your casualties are and if there is any way you can get outta there. Can you do that for me kid?"

"Yes mike, I'll see what I can do and get back to you."

I turned to the guy across the path on the right and explained the situation. I asked him to try and spread the word down quietly. I didn't think it wise to yell our plans out. Word came back that casualties were heavy and we would not be able to leave.

I waited a few minutes before contacting the CO. I had to take a breather. I looked at my watch. It was almost six.

I stared at the body again. Who is, was that? I had been in this spot for three hours now and still didn't know. I looked at the guy on the right. I wanted to ask him who it was, but he was lying back with his eyes closed probably thinking as I was. I decided to leave him to his thoughts.

I stared at the body some more. The portion of the face that I could see had turned an even darker shade of yellow. It had stiffened up, probably rigor mortis had set in. It no longer lay there in the casual pose I had first seen. Of course the flies were everywhere. Damn those flies. Why couldn't they leave him alone?

I wondered how many others were wounded. I felt that we were fortunate to only have one dead. It could have been a lot worse.

"Mike, this is xray."

"Go ahead kid."

"The lieutenant says no. Casualties are heavy and we are not able to leave."

"Roger. Okay kid gimme the list of casualties and their seriousness."

"Roger, wait one."

"Sir he wants a list and seriousness."

I was yelling at the lieutenant. It didn't make any difference anymore. We obviously did not have any plans to keep secret. The guy on the right perked up and was watching me. The lieutenant didn't answer. I yelled again.

"Sir, the CO wants a list of casualties and their seriousness. Sir… can you hear me?"

"Yes. Four KIA and three WIA."

"What the fuck is a k-i-a and w-i-a."

The guy on the right answered.

"Killed In Action and Wounded In Action."

I stared at him in horror. He repeated it.

"Four … killed and three wounded."

I kept staring. I couldn't believe my ears. We had four men killed. There were three others besides the one I had been staring at. There was an eerie silence now. Most of us further up had no idea of how bad we had been hit. I contacted the CO.

"Mike, over."

"Yes, go ahead."

"Mike we got four Kilo India Alpha and three Whiskey India Alpha."

I used the words for the letters so they would be clear.

"Repeat that kid."

"Four dead, three wounded."

"How bad are the wounded? I don't think we can get another dust off in."

"Wait one, I'll find out."

"Sir, how bad are the wounded?"

"Hold on kid I'm gonna try to come up."

"Mike the lieutenant's coming up to me."

"Roger kid."

I rolled over on my back and waited for the lieutenant to arrive. I still couldn't believe it, four dead. Who else had been killed? Hell who was this? Those bastards up there would pay for this.

The lieutenant arrived and slid in next to me. We hadn't been fired at since the last dust off had left. If only we could get out of here. The lieutenant stared at me for a long time before he spoke.

"You did a hell of a job kid. I'll see you get a medal for this. Ya know ya got the only working radio in the whole platoon. Smitty's got busted up when he got hit…"

"Smitty got hit? Is he okay? I mean…"

"I know what you mean. Yeah he's all right. Got it in the thigh went clean through. Hit the bone though, probably a good fracture, but he'll be okay."

"That's Jeff there never knew what hit him. Got it right in the head. First one hit. Benny's down a little ways. He panicked when Jeff got hit and froze. They just shot him to pieces. Tom's down a little further. Shot him right between the eyes. His glasses spit in two. Probably never knew what hit him either. Guam's down on the LZ. They shot his face up. Just kept shooting. I don't know why. After that they just stopped."

"Right after the dust off pulled out we were moving him back off the LZ, Ray was holding his head and shoulders. All of a sudden they opened up. Shot up Ray's arm pretty bad. Medic says it severed an artery. Says if we don't get him out tonight he'll lose the arm. When he got shot he dropped Guam. That's when they started shooting Guam. Just kept shooting. Damn. They had to know he was dead. I really don't understand it."

"The rat's gone stone deaf. Musta been from all that shooting he was doing. The professor's got him in tow." The lieutenant pointed to professor's group just above us.

"I'm gonna try to get everybody down to the LZ right after I talk to the CO. We'll have to spend the night here. Take a breather kid. You deserve it. Ya did a helluva job."

"Mike, this is xray, over."

"Thank you sir. I appreciate that."

I lay back while the lieutenant informed the CO of our situation. He would have to round up the pack numbers for those killed and wounded. We would not use names on the radio. We had all been assigned pack numbers to identify us for one reason or another so that the CO could keep a record of who was out in the boonies at all times.

I could almost see the other platoons spreading the word about who was killed as the lieutenant read the numbers. Although we hardly ever worked with the other two platoons we had still heard of and knew each other. It would affect them too having lost four buddies.

It was one thing to lose a friend when he left the boonies and headed back to the world, as we called the states, but when he was killed, you instantly felt the permanence of it. We had all talked amongst ourselves about our friends and loved ones back in the world. We knew how they would feel, just as we knew how our own would feel if we didn't come back.

There was a very close bond between us all regardless of religion or beliefs or race. We all had one objective and that was to stay alive. We would cross any barrier to achieve that goal. It wasn't just a friend we lost. It was a part of each of us.

The lieutenant handed me the radio.

"Kid, guard that with your life. It's the only one we have left."

"Yes, sir. What do we do now?"

"Let's get everyone down to the LZ and organize from there. Try to stay as quiet as possible. Let's go."

With that, I motioned to the guys on the right to start moving down. The lieutenant went up to get the professor and his group. I waited while the professor's group moved past and then started behind the lieutenant. One guy stayed behind me to cover our exit.

The path was only about two feet wide and twisted around large rocks. This was the first time I had been on the path, except when I came down. I crawled around Jeff's body. We had to leave it and continued downward.

I passed Benny, his body had been charred when the LZ caught fire, not badly, but enough to leave a black crust over his chest and face. The clothes had been ripped off and were strewn to the side. That had been done in a frantic effort to keep the body from burning.

Around the bend and down a little further lay Tom's body. The face had been covered over with his shirt and a great deal of blood had soaked through. His glasses were off to the side, broken in two at the nose brace and each lens cracked from there to the outer edge.

We would leave these two as well. We continued our trek downward. I had not realized they were so far down. The LZ was just around the next bend past another set of huge rocks. We filed in quietly and filtered into various positions.

Guam's body was at the opening of the LZ. The face was covered with his shirt and also soaked in blood. His pants were pulled down around the ankles and there was a gaping hole in his abdomen where he had been initially shot. We had to leave the body like that. We had nothing else to cover it with.

Smitty was in the middle, his leg heavily bandaged. Next to him was Ray. A tourniquet wrapped tightly around his right arm. He was lying quietly. The medic had pumped him full of morphine. The professor steered the rat over next to Ray. We would keep our three wounded in the middle of whatever kind of perimeter we could devise for the night.

We traveled about one hundred yards to get here and it was the longest hundred I had ever traveled in Vietnam.

I moved in toward the middle and set up my radio next to the wounded. Sarge came over and shook my hand complimenting me on the job I had done. Said he would talk to me in the morning. The lieutenant came over took the horn from me and reported our progress to the CO.

It was dark now. We were all quietly laying down keeping a watchful eye on whatever might happen. We had left all of our gear up in that rock formation in our hasty retreat. The gooks were probably enjoying their capturing of so much GI gear that they weren't worrying about where we went. At least we hoped so.

We had left weapons, ammo, food, water, helmets and various personal items behind. All we had was what we could carry with us. No one knew this was how we would wind up and could not be selective of what they brought with them. I suppose each of us thought about that as we lay there.

We would have to spend the night, the morning and probably the next day without food and water. But most of all our ammo was very low. We had done a lot of firing.

I lay there past midnight waiting for something to happen. The tension was tremendous. Everyone was edgy. If the enemy wanted to they could really do us in. We were basically defenseless. For some reason we were not bothered. We waited and waited but nothing happened. Morning would come soon. We would attempt another dust off. A little while longer.

As daylight broke, we gathered ourselves together and prepared to get out of this place. We would have to bring the dust off in first. We could not carry out the wounded. I made the call and we waited for it to arrive.

We continued to remain very quiet. We did not want to announce our presence in case the enemy was still near by. We knew they were, but maybe we could get this dust off in before they had a chance to react. The radio became active.

"Xray, this is dust off one niner. ETA approximately five minutes. Do you copy? Over."

"Roger one niner, smoke going out over."

"Roger xray. I'll come out of the valley over."

"Roger one niner we'll be ready."

We popped a smoke grenade, one of the few we had left, and waited for the helicopter to drift in. He came up in a flash just like the last one the day before. As he hovered about four feet from the ground, we scrambled to get the wounded aboard.

Just as we were pushing the last man on a blast of gunfire exploded around us. The helicopter rocked, but was not leaving. The pilot held it in place until we got the last man on. Just as he had arrived he was gone. I crawled back over to my radio and tried to reach him.

"Dust off one niner, this is xray over."

"This is one niner over."

"Roger one niner, are you okay?"

"Roger xray. We took some hits, but I think we'll make it back okay."

"Roger one niner… and thanks."

"Roger xray. Glad I could help."

"One niner. I just wanted you to know that we appreciate what you did for us. It took a lotta balls to come back out here after yesterday. Thanks again."

"Roger xray. I get your message, but you still got to get outta there man. I'm going back now. Wouldn't want to trade places with you guys down there."

"I don't know one niner. You guys are sitting ducks flying into hot areas like this. I think I'd rather be on the ground."

"Roger xray. You got a point, but right now you're in the hot seat, so keep your asses down and get the hell outta there… and xray, if you ever get back this way, stop in and I'll buy you a drink."

"You got a deal one niner… take care…"

"Roger xray…"

# THE PROTECTED WILL NEVER KNOW

In an instant the helicopter was gone. We had got the wounded out. We would start making our way down and let the big guns take care of this fucking hill. We had paid our dues.

Well Jack, your percentages held.

# ESCAPE

**WE WAITED AWHILE** to find out what the enemy would do next. They had fired at the helicopter. Would they attack us again? We had to be sure before we started to leave.

After a short time, when no other shots were fired, we decided to make our move. We quietly gathered the little gear we had with us and prepared to move out.

The strong wind that had been generated by the helicopter's hovering had blown things about, including the shirt covering Guam's face. Actually there wasn't a face anymore just a mass of torn tissue that created a grotesque picture of one man's inhumanity to another. One shot would have been enough.

Why such a vengeance to Guam and not any of the others? It was a question we all needed an answer to. One of the guys nearest to his body retrieved the shirt and placed it back over Guam's face.

We would have to leave the body there, as well as the others. We had no choice, but someway, somehow we would be back to get them. No matter what it took. We vowed that before leaving.

## THE PROTECTED WILL NEVER KNOW

As a bond to our vow, one of the guys picked up the poncho liner that we used to cover Smitty and Ray with during the night and started ripping strips off of it. We had entered this hill with twenty-five men, but we were leaving with eighteen. Sections of the poncho liner were covered with blood, but that did not matter. Each of us took a piece and tied it around our necks. We named those strips our "drive on rags". Each guy that wore it would somehow see to it that the bodies were recovered and the score was settled between us and hill 474. It was a promise that we would keep even if it were to the last man.

We had two things going for us while getting out of there. We would be going down hill and we would not have to carry the heavy gear we normally did. After all, we had donated all of it to the gooks.

The path that lead to the LZ also continued down the hill as well. It was apparently a path throughout the hill. We would follow it at least for a little way. It would be the quickest way out. It followed parallel for awhile then dropped down a bit, then back up. It was taking us along the side more than it was taking us down.

After awhile, the lieutenant called, the CO to let him know our progress and receive further instructions. The CO informed us that there was a ridge a few meters down and further to the right, away from the hill. We would actually be on another hill next to 474.

We were to get to that hill immediately and wait for supplies, basically food, water and ammo. The lieutenant informed the CO that we would not be able to get there right away. We would have to cross a ravine first and that would take some time.

Apparently the CO was not impressed. In the next instant the lieutenant gave me back the handset and said to move out fast. As we got up to leave he explained that heavy artillery was scheduled to start shelling the hill any minute and unless we wanted to be in the middle of it, we had better move quickly.

We got off the path now. The fastest way to get there was straight down and over. We headed that way. Near the bottom we approached a stream. We would have to go across it.

Upon entering the water, we discovered it sloped down as we crossed, creating a steep shore on the other side. It would not be easy to get back out. We could not wait. We stayed in the water. It led to an opening in the hill that was actually a patch of overgrown vines, giving the stream a protective covering.

The point, or lead man, followed it in to find out how far it led. He returned in a few minutes explaining that it came out a few meters down and it appeared safe enough.

In some spots the water went from knee deep to waist deep, but the water felt refreshing this time. It cooled us off from the forced marching we were doing.

We were happy to discover it brought us within a few meters of our desired spot. We could already hear the shells starting to pound the hill above us.

Upon arriving at our destination we discovered that the ridge had two sides with straight drops of about forty feet and the other two were an uphill climb to reach the top. This would be safe enough. However, we thought that last time too.

We contacted the CO and informed him of our arrival on the ridge. The CO instructed us to wait for the helicopter, grab a bite to eat and make our way to the bottom. We would be met there.

In a few minutes, the pilot of our re-supply helicopter announced his arrival. He informed me that he had spotted us and we would not have to pop smoke for him. I wasn't sure if we had any smoke grenades left to pop.

## THE PROTECTED WILL NEVER KNOW

We noticed four gunships circling the area around us. They would fly in close to the side of the hill and hover as if they were daring someone to challenge them.

We were resting a little easier, having all that protection. The shelling of the hill was becoming heavier. We gathered by the time lapse between the explosion and the sound that we had covered quite a distance.

Whenever a helicopter landed out in the boonies, it would stay at full idle just in case it would have to make a quick getaway. As was the custom, one of the men would direct the pilot to let him know when the skids had touched the ground. With this being a re-supply drop, we wanted it flat on the ground so it would be easier to unload.

Jack directed the helicopter in this time. As soon as he indicated to the pilot that he was down, Jack raced to the co-pilot's door and opened it. He was trying to tell the co-pilot something. From the distance I could not tell what they were talking about.

There were two men on each side unloading the cases. The men were just grabbing the supplies from the helicopter and throwing them to the ground. Even if it was safe, we still tried not to keep a helicopter sitting too long.

Jack was still talking to the pilot and co-pilot. Actually it looked more like he was arguing with them. They would be done unloading soon and whatever it was Jack wanted would have to be taken care of soon.

I was watching the last of the cases being taken off and heard the pilot announce over the horn that he was preparing to lift off. I looked toward the pilot and Jack was still there.

Finally they handed Jack something and he backed off and closed the door. Jack stepped back and the helicopter lifted and took off. I watched as Jack stuffed something into his pocket and started walking toward the cases lying on the ground.

The cases contained c-rats (c-rations) and ammunition. Even though it had been almost twenty-four hours since we had eaten, food was not greeted with enthusiasm. Everyone was finally getting a chance to stop and think about what had happened the day before and food would not settle too easily in our stomachs.

I was busy working on my radio. They had sent a new battery, so that I could change it. As a rule, I would change my battery every three days, but the realization that this was the only working radio we had left was now beginning to sink in.

I had called the CO and informed him that I would be off the air for a few minutes while I changed the battery and cleaned it up a bit. The radio had been through a lot in the last twenty-four hours. I wanted it back in the good shape I kept it. I had trouble with the radio a little while before we made contact and I wanted to be certain the repairs I had made then were still holding. As I was doing this, Jack came over and sat down by me.

"What's happening kid?"

"Hey Jack. I'm here."

"I can see that. How's the horn holding up?"

"It'll make it. Say Jack, what the hell were you hassling those pilots about?"

"Stuff man. I left mine in my ruck. I couldn't see going through another day without it."

"You mean they gave it to you just like that?"

"Christ no. You seen what I hada go through ta get it. I didn't have no money so I hada promise I'd get them a souvenir from the hill."

"What kinda souvenir?"

"How the fuck should I know? Something gook. Whatever I can find I'll send them. They gave me their unit and when I get something I'll send it to them."

"Is that what I saw you stuffing into your pockets?"

"Yeah. Boy you sure are observant. I thought I did that pretty well."

"Not really Jack. If I saw you, then I'm sure someone else did. What if the lieutenant asks what it was?"

"Fuck him. Let him get his own shit. I'm not giving this stuff away. Fuck man, you sure are paranoid."

"Not really. I just thought I'd ask. Where the hell did you disappear to yesterday? I didn't see you at all. Did you get stuck down on the LZ too?"

"Yeah, shit. I was down there. Smitty, me and the sarge went down first to see if we could secure it. Next thing I know, Smitty screams and drops to the ground. I ran back to him and he's rolling on the ground holding his leg. It's bleeding something fierce, so I called for the medic and the sarge. He's maybe twenty feet behind me. He tells me the medic is coming, but then Benny starts screaming and pretty soon he ain't screaming no more. It got real quiet for a minute and sarge picked up the radio. I guess that's when he called you, but the radio was busted. The handset cracked in half. Sarge put it back together and talked for awhile, but when that other bird got hit, he dropped it and couldn't get it to work no more. The medic came over to Smitty and bandaged his leg and gave him a hit of morphine. I don't know what happened to Tom. I'm not sure when he got hit. It was some bad shit kid. It all happened so fast."

"What about Guam?"

"You sure you want to hear that kid?"

"Jack, I'm not a cherry boy anymore, remember? What happened?"

"Yeah, that's for sure. Think we got time to smoke one?"

"Yeah, I think so. You got anything to roll it in?"

"No, got a pipe. Always carry it just in case. Ever since we were in that fuckin' rain."

"Wait until I tell them I'm back on the air."

Jack removed the package from his pocket and preceded to stuff some into the pipe he was carrying. I let them know I was transmitting again and would be monitoring the traffic. I watched Jack use his lighter to start the smoke.

"Here kid. It's ready, take a hit... take another... another..."

"Jack..."

"Trust me kid you'll need it... another... okay."

I gave the pipe back to Jack and leaned back with the handset on my shoulder next to my ear. Jack took a heavy draw, then another, finally he spoke.

"I really don't know exactly, but it was quiet for a long time then you yelled to pop smoke. That's when the fuckin' thing exploded and set the grass on fire."

"We had Guam hidden behind some rocks. He was doing okay, but when the fire started we had to move him out in the open. That's when the gooks opened up. I heard Ray scream and fall to the ground. Ray had fallen behind Guam and was rolling from side to side. Hell, the gooks coulda got Ray too, but they just wanted Guam. The gooks shot him five, six times. God only knows why. After that, the gooks just stopped. Some strange shit. I'm telling ya we got our asses kicked. Ain't never been in no shit like dat before."

"Damn Jack, I wonder why they had it in for Guam. Lieutenant said the same thing. Just kept shooting."

"The only thing I can figure kid, is the gooks thought he was Vietnamese, being from Guam he looked kinda like them. The gooks musta figured he was one of them, come over to our side. Other than that, I don't know why they had it in for him."

We sat there quietly puffing on the pipe and staring off into space. The "stuff" as Jack called it was really helping ease some of the tension I had. I was feeling more relaxed now.

Dying was a fact of life and you just had to get used to it. Although, I still had visions of what I saw on the path that day and probably would for some time, I was beginning to accept it. You had to in order to continue existing over here. I was beginning to wonder what the next ten months had in store for me then I remembered Jack's percentages.

"Well, Jack, you were right about me."

"Yeah kid, everyone else too."

"What?"

"Smitty and Jeff. I told Smitty he'd only last two months, but not to worry. Jeff, I had bad vibes on. He wouldn't believe me, thought I was bullshitting. Nothing he coulda done though. They don't pull a guy outta the boonies on hunches."

"What about the rest?"

"Didn't know the rest. Didn't really know Jeff, but we got ta talking a couple of days ago when you was in the rear and I had a feeling his time was up. It's weird man. I didn't worry about you though, knew you'd be alright."

"Wish you woulda told me. I wouldn't have been so scared."

"Couldn't. You woulda fucked up then. Had to be the way it was so's you'd act the way you did. Understand what I mean?"

"Yeah I see what ya mean. Hey, we better be gettin' on."

"Yeah... want anymore?"

"No. I still have to walk down. I better lay off."

"Okay, see ya later kid. I'm walkin' point."

"Yeah, see ya Jack."

We were actually quite close to the bottom. It only took us about an hour to get the rest of the way down. We came out around the other side and walked closer to the base of 474.

The bottom was mostly flat land, with a lot of rice paddies. There was a lot of activity going on down there.

Two helicopters were off on a small incline. Their engines were off and they were just parked on the ground. Apparently they were two of the dust offs yesterday that couldn't make it back to English.

There was also a unit of the 173rd waiting for us to move into place before they moved out. They had some food and water, cigarettes and mail for us that had been sent to them to give to us when we joined up.

We set up and dug into the mail. Mail was probably the most important thing a grunt could receive over there. It was always a great morale booster. We could really use some morale boosting about now.

We would be staying a couple of hours, then we would be flown back into the rear for the night. We had done our job for awhile. We could rest a bit. Our spirits were up.

We watched as the bigger Chinook helicopters came out to pick up the disabled smaller ones. Another unit of the 173rd came into the area to guard while the disabled helicopters were taken away. These bigger helicopters were twice as big and were commonly known as "shit hooks". They were primarily used to carry things in and out of areas as well as move troops.

A few of the guys from the 173rd wandered over and began chatting with us. They informed us that 474 was completely surrounded now and they would catch anything being forced out from the pounding the artillery was performing on the hill. The plan was to shell the hill continuously for the next week. Tomorrow they were bringing F4 fighter jets in to bomb the hill all day long. They were also going to drop fugass (an explosive gas carried in 55 gallon drums with an extremely flammable quality) to burn out all of the vegetation and penetrate the caves.

In the next week we would own this hill and everything in it. They didn't know yet what we had walked into, but by the end of the week we would all find out.

We had accomplished our mission. We had been sent to seek out the enemy and report its position. That we had done. The rest would be left to whatever the higher ups could drum up to uproot and destroy the enemy.

That was the name of the game. In terms of losses we had paid a very small price for this information. In our eyes, any price was too high, but that was the past. Now we were the victors. As for this hill it no longer existed as an enemy fortress.

Jack and I smoked some more. In fact, Jack was feeling better now and decided to share his "stuff". The guys from the 173rd had re-supplied us with enough to go around, besides we would be spending the night in the rear. We could relax now.

The helicopters started arriving to take us out. Since there were only eighteen of us and it wasn't a combat assault we piled into three helicopters. I watched the valley drop below us and the hill disappear behind me and wondered if it looked like that to the medivac helicopters as they pulled away. Unfortunately, they weren't leaving as peacefully as we were. I laid back and enjoyed the ride in.

## COLLECTING MY THOUGHTS

THE REAR WAS A BIT FRIENDLIER THIS TIME. The last time I was here I was alone. I could be hassled easier. Now we were here in force. Our whole platoon, or what was left of it, came back in.

It would not be so easy to push all of us around. Besides we kept our rifles. They had requested that we unchamber the round, as well as disengage the magazines, but we still had them in our possession. It made our position known.

However they were not out to hassle us this time. In fact, the REMFs were eager to comfort us. There was a hot meal waiting, clean clothes, new gear and a shower with hot (well warm, anyway) water.

We would only be here until morning, but it would be relaxing. They did not even comment about our unshaven faces or the various artifacts we had hanging about ourselves. Maybe there was hope.

The rear, as it was called, had been moved since the last time I was there. Our unit had two tents set up near the 173rd headquarters section, which I suppose added to the pressure of keeping things orderly.

Since then, three tents had been set up further out of the central headquarters section, nearer to the helicopter strip. The guys that ran our rear had said that we were moved in order to get, "... the scum away from the brass..." The 173rd had explained the move as a chance to set up a better location with three tents now. We all tended to believe our guys.

Of course, it put us away from everything, including the mess hall and the souvenir shop. The area did have its advantages, in that it was away from any close scrutiny so we were able to do as we pleased.

I had hit the souvenir shop for a new Zippo lighter. As was custom we had our unit markings, company, battalion and such engraved on one side. On the other side I had engraved:

**Vietnam '69-'70** on the lid and:

**"For those who fight for it, life has a flavor the protected will never know,"** on the base.

There were many quotes and sayings that were floating about and guys had them on lighters and jackets and on helmets and just about everywhere else. Of all the various sayings I liked that one the best, because it said it all.

The second best for me was a variation on the 23 psalm; "Yea though I walk through the valley of death, I fear no evil for I am the evilest son-of-a-bitch in the valley." I heard and saw that phase in various forms but this version was the most common.

The most prevalent symbol or expression was the peace symbol. We all wore it in some form or another.

After showering and changing clothes, we started the long trek to the mess hall. Top (the first sergeant), who ran things in the rear, had convinced us to leave our rifles back at the tents and not carry them with us through the compound. We had argued that we might need them, but Top indicated the only enemy we were likely to meet would be guys from the 173rd and they weren't killing enemies, they were beating up kind. Hell, with that kind of logic he had us, so we started our trek defenseless.

This was my first "sitting down at a table, eating with silverware" meal in a long time and needless to say it felt kind of strange.

One thing I can say about the mess halls in Vietnam was, there was no restriction on how much you could eat. You could keep going back for as many helpings as you wanted. Food was definitely plentiful. We ate until we stuffed ourselves.

Afterward, we sat around smoking (regular cigarettes) and joking. We didn't even have to clean up our mess. That was done for us by a group of GIs whose permanent assignment was mess hall duty.

I didn't quite understand how any grunt, especially one in Vietnam would agree to perform permanent KP (kitchen police) duty. One of the guys explained that he had heard about a special group that would normally be too dumb to get into the army, so they devised a plan to let these people in to perform special duties, such as KP and other meaningless shit duties. I wondered just how dumb you had to be to be too dumb to be in the army.

Anyway, they would have a hell of a time forcing someone like us to do it. Hell it was great for us. The more I thought about it, the more I went back to Jack's theory of further hassling you in Vietnam. It all made sense. We decided to start heading back before it got too dark to find our tents.

When we got outside we realized we had been in there too long. It was already quite dark. We had more or less broken into small groups and were milling about outside the mess hall trying to decide what to do next.

As long as it was dark already, there was no need to hurry back, although some did go back. The rest of us headed for a Quonset hut that had a movie showing inside. I was just getting into the movie when Jack decided we should leave.

"Com'on kid, let's go."

"Wait Jack, this looks good."

"Fuck da movie, let's go grab a smoke."

"Why don't we smoke around back and see the end."

"What are you? Some kinda fuckin' movie freak? For Chris sakes, ya can read the damn book, shit."

"Jesus Jack, if you're gonna get all fucked up about it we'll leave. Boy you sure are a prick when you're down. Are you that dependent on being high?"

"Ya wanna go or do you wanna lecture?"

"Alright, we'll go... shit..."

Jack and I left the Quonset hut and started to head back to our tents. We walked along in silence most of the way back.

When we got to the tents, the spare tent was already quite full. Most of the guys that came back were already asleep and there were sets of gear to reserve spots for those that were still at the movie.

Jack and I had not taken the time to reserve one. We came to the conclusion that it would be better to sleep outside anyway. It would get quite stuffy in there. We grabbed our gear and headed to a spot a little ways off to the side of the tents. We got settled in and Jack brought out his "stuff".

"Man, look at dem stars. Ain't it beautiful?"

"Yeah Jack, it sure is. You know, it sure is nice relaxing for a change."

"You ain't relaxing kid. You're just as tight as ever, probably more so. It just feels like you're relaxing, 'cause you ain't out in the boonies. Just look at your hands, ya still got the shakes and ya damn near smoked a full pack a smokes in the last coupla hours. How many packs a day you smokin' now?"

"Hell, I dunno, four maybe five. So what, afraid I'll get lung cancer?"

"No man, don't give a shit if you do. Just makin a comment on your relaxing. If you was dat relaxed you probably wouldn't be lightin' one off the other. Here, put out the damn cigarette and smoke some of this pipe."

"How come you got the pipe?"

"Less hassle. We won't have to stop so often. Christ, I gotta explain everything to you. Don't you know nuthin?"

"Hurry up and smoke will ya, so you get mellower. Don't talk no more until ya do… shit…"

Jack and I lay there quietly, just drawing on the pipe waiting for it to happen. We had loaded the second pipe before we decided to chance conversation again.

"Hey kid, ever think about home? I mean what you're gonna do when you get back?"

"That depends on how… and when… I get back…"

"Will you forget that shit for awhile. Christ ya gotta make an issue outta everything… shit… I used to be able to talk to ya… now Christ… let's just say you do your time and ya get sent back to the world… then what? Can you answer a simple question? …shit"

"Hey yeah… I'm sorry Jack… I was just fuckin' with you… for one thing, I was considering extending so I wouldn't have to do no stateside duty…"

"Yeah that's cool… yeah, well assuming either one, what are your plans? …huh?"

"Well… if I don't extend, I'll try to just serve out my time quietly somewhere, so I don't get fucked over like you did, but it was for that reason that I was considering… you know… extending… so I could go back to… ahhh… you know… ahhh… what da fuck were we talking about?"

"Extending…"

"Oh yeah… I'll extend to avoid stateside time."

"Then what?"

"What?"

"Then what are you gonna do?"

"I'm not sure. I got me this chick back there… we'll probably get married eventually… but one thing for sure, we're gonna have a blast makin' up for lost time… I'm wasting' some prime years fuckin' around here… you know, she's really an okay chick… writes me alla time… keeps me informed on what's happening back in the world… I really dig her a lot… I wish I was with her now… shit I hate this fuckin' place… and this whole fuckin' army bit… it really sucks… it really, really sucks… I want outta here…"

"Hey, take it easy kid… you'll get back there… ain't no use gettin' worked up, you can't change nuthin… stuck like the rest of us… what ya gonna do about it… fool around and get busted, you'll just have to serve more time… fuck it… it don't mean nothing… nobody cares… to them here you're just another piece of meat, back in the world, you're shit for being here… just take it a day at a time… keep thinkin' about your chick and the good times you're gonna have…"

"You got anybody waiting Jack?"

"Yeah... nah... well maybe. Had this chick I was foolin' around with, but we kinda split before I left... heard she was knocked up... It could be mine I guess... we weren't too serious, just had some good times... might look her up when I get back... nobody really, just some family... Is that a shooting star or a flare?"

"Whatever you want it to be... got anymore? Pipe's going out."

"Yeah got lots... here..."

"Wow, how much of this stuff you got?"

"Again with the fuckin' questions... you a spy or something? Christ... don't worry, I got a continuous supply... shit..."

"No... just curious... Christ are you touchy... you always been such a prick?"

"Only when I'm getting' the third degree all time... anything else you want to know?"

"I'm really sorry Jack. I guess I do ask a lotta questions... but you're probably the only real friend I have here... and... well... shit I don't know... I guess I just am trying to find out what it's all about... and ... ahhh... you know... you been here before, so I figured you knew all about it so... ah fuck it, if it bugs you that much, I won't ask no more questions... I'm sorry man... really sorry... so forget it okay? ...just forget it."

"Whoa, slow down kid. Ain't nuthin to get riled up about... I just get tired of getting' the third degree alla time... hey shit, just gimme a break once in awhile and I'll be okay... com'on Kid what were you talkin' about?"

"Who cares?"

"Now who's the prick?"

With that we started laughing hysterically. There was no need for us to get on each other's back. We had enough problems without turning on one another.

We lay there silently for awhile, just drawing on our third pipe, neither one of us fully conscious of what was happening around us. It was peaceful. I would sleep good tonight. So would Jack for that matter. Tomorrow we would go back out to Hill 474 to finish what we had started.

I awoke to a blinding sun. I looked at my watch. It was eight-thirty am. God, I had slept in. I hadn't slept this late since I got here. It was kind of nice. I looked over at Jack. He was still asleep. I decided to risk whatever would happen and wake him. He was awake instantly and staring at me with those same burning eyes he had used on the Chaplain. After he regained his bearing, he relaxed and rolled over into the sun.

"Christ, that's bright."

"It's alright once you get used to it. We're not use to getting up in the sun. Let's go get some chow."

"Yeah. What time are we pulling out?"

"Coupla of hours I guess. I heard them say we wouldn't be here for noon chow."

"Shit, can't wait to get rid of us."

"Jack, about last night…"

"Forget it kid, it's history."

"Yeah, but…"

"You gonna start or are we gonna eat?"

"Let's eat."

We went and ate breakfast. It would probably be the last regular meal we would have for awhile, so we made the best of it.

When we got back to the tents, we busied ourselves getting our gear together and preparing to go back out into the boonies. I had built a pretty good ruck the last time I was here, but this one was even better. It helped having all new equipment. The last time I had to scrounge around for pieces.

We all had to write our friends and relatives and whoever else might have written us to explain that our gear had been captured and with it some letters. It was possible that the enemy might write them and feed them some bullshit about us being captured or something. The lieutenant made sure everyone wrote one.

We also had to write each other up for medals. We composed the details as best as we knew of what others were doing during the firefight. They would take these summaries and submit a recommendation for a medal.

One thing I discovered about medals was that the most decorated soldiers, that is of the lower ranks, were clerks. Since they did the final paperwork and submitted the requests it was easy for them to submit their own. Interestingly, it really placed the true value on a medal.

Will the real enemy please stand up?"

We sat around waiting to depart. Some new cherries (recruits) had joined our platoon and were being shown the ropes. The fellow who had been stuck carrying the M60 machine-gun instantly handed it over to a new cherry boy. The extra radio we carried, actually Smitty's radio, was also passed over to another cherry boy.

As I watched all of this transferring take place, I thought about giving my radio to someone. After all, I had taken my turn. For various reasons, I decided to keep it. It had done well by me so far. Maybe it was instrumental in saving my life. A little far fetched I know, but it was a good justification for hanging onto it.

## THE PROTECTED WILL NEVER KNOW

As I sat there waiting, I thought back over the last two months I had spent here in Vietnam. I surmised that in those two months, I had been through just about everything you can go through in Vietnam.

I had shoveled shit, been to the house, built a fire base, slept in the mud and rain, fallen off a cliff, been in a firefight, seen four guys killed in action and ran for my life. Oh yes, I was turned on to smoking.

Not too bad for two months work. The one thing that intrigued me about all this was the "so what, that's life" attitude I was developing about everything. If all of this had happened in just two months, what would the next ten be like? Or if I believed Jack, the next four.

We would be going back to Hill 474. Would we make contact again?" Would I survive another? Would I repeat the cycle and go through everything again? Damn I hated this place. There was absolutely no way of planning the future. Over here there was no future. There was only the present and the past. That was sometimes doubtful.

How can you reasonably exist not knowing what is going to happen tomorrow or even the next hour, or next minute, for that matter? I was beginning to understand why guys blew their mind over here.

This was not a real war. That is, not a war as we were taught a war was supposed to be. This was a skirmish. Sometimes you run into what is interpreted as the enemy and sometimes you think you do. Do you let them all go? Or do you kill then all? If you let them go, they will probably kill you. If you kill them all, then you are a monster or a murderer. Damned if you do, damned if you don't. Fuck it… it don't mean nuthin.

Will the real enemy please stand up?'

You are smack dab in the middle of it all, trying to get by day after day. "You're here to follow orders soldier…" is the bullshit you have been told since the day you arrived, since you entered the army for that matter.

You are ordered to enter a village and destroy it, including its inhabitants, you destroy it. Word gets back to the world about what you did and you are condemned for it. It doesn't matter that you saw suspected VC in the area and some of your buddies have been killed or wounded, you shouldn't assume they are all VC.

Just because you captured a fourteen-year old boy running guns out of the village into the hills or you stopped a group of women coming out of the hills, that turned out to be mostly men carrying supplies. Hell that's no reason to get suspicious. The village is still friendly. Don't they always give us the bodies back?

A person can only take so much regardless of what the consensus are led to believe. One day it happens. An accidental bombing run, or a barrage of artillery off course, or gallons of fugass misdirected. Oops sorry, bring out the engineers and bury it. Just cover it over. The jungle will grow over the ground. In a few days, no one will know. What village? There's no village there. You must have wrong information.

The sad fact is that those who knew didn't even care, nor did they care what others thought. I was beginning to feel the same way. As Jack had said last night, "Fuck it… it don't mean nuthin…" Who was I to carry the burden? Why should I worry about anything? My only objective was to stay alive. That I would do at all costs.

The guys were starting to stir. We were moving out. I gathered my gear and joined the group. We loaded into trucks that would take us to the helicopter pad, for transport back to the hill.

## THE PROTECTED WILL NEVER KNOW

Just as I was getting settled, the truck jerked and I and my one hundred pounds of gear fell backwards. I extended my hand to break my fall and watched my right thumb bend the way it shouldn't. After a barrage of cursing, I let the medic look at my hand.

The medic suggested I get off the truck to stay back and have them look at my hand. I asked if anything was broken and he said no just a nasty sprain. I asked if it could be bandaged enough to let me go with the platoon and he asked if I wouldn't rather stay here. I shook my head no as well as said it. I was not staying here alone. Besides it was my right hand and I used my left to shoot.

We boarded the helicopters for the flight out to the hill. We could relax a little. We were landing in a secure LZ. This would not be a combat assault, at least not a true CA.

I had cleared my mind of my earlier thoughts and was concentrating on my aching hand. I watched as the helicopter pulled up and away from the pad. It was back to business.

# BACK TO 474

**THE TOP OF 474,** although it peaked to a point, was actually quite large. There was enough room to bring a helicopter down for a landing. We landed one at a time and I was right, it was a secure LZ. Our second platoon was camped there. The helicopter could actually land flat. It would not have to hover at full idle as it usually did. The ability to step off instead of jumping from the skids was a welcome relief.

We milled about until the last bird landed, unloaded its passengers and took off again. Then we found spots and dropped our gear. We mingled with the guys from the second platoon and exchanged stories about the contact. The firefight was the first for quite a few of them as well.

After we broke for noon chow, we sat around listening to the CO explain the agenda for the coming weeks.

The first week we would move off the top, down the other side to a smaller peak and wait while air strikes (bombing runs) would be leveled on the side of the enemy positions. At night, we would move back to the top and watch while artillery would be leveled against the side of the hill.

## | 124 | THE PROTECTED WILL NEVER KNOW

The second week we would start making the sweep down, in company strength, to search out any remaining enemy and collect anything they had left behind.

On the bottom of the hill were elements of ARVNs (Army of the Republic of Vietnam) to catch anyone trying to escape that way. The ARVNs were the locals that were mostly youngsters (hell, so were we). They were the South's (South Vietnam's) fighting force and believe me as a whole they left a lot to be desired as a fighting force. Their disorganization caused them, as well as us, many problems. We were glad they were all the way on the bottom away from us.

On each of the sides of the hill were elements of the 173rd to try and stop anyone trying to escape that way. They particularly guarded the southern escape route into a nearby village. Like most villages, the people were sympathetic to whoever was in the neighborhood at the time. The village had been supportive of the NVA-VC, but now they were friendly to us. We let it be.

The backside of 474 lead to another smaller hill. In fact, it was the same hill we would move back to every day, so we would leave one of our platoons there every night as a guard and lookout post. In all actuality, the enemy would have to come over the top to get to it and that was unlikely. Mainly what we were after was the one or two stragglers that might wander into the area.

I suppose you could honestly say we owned this hill. As with all other hills, once we finished with this one, we would give it back to whoever wanted it. For now, hill 474 was ours and we would do with it as we pleased.

Today we would remain on top. We would start our daily trek down tomorrow. We mostly milled around and generally wasted time. We explained, as best we could, to the new guys what had happened and why we were here. They were impressed and scared at the same time. The new guys did not expect to be in the thick of it their first time out. I could sympathize with them. I had not run into anything until my second month. However, it was just as well to get it over with now, instead of waiting and acquiring a feeling of relaxation as I did.

The next morning, we moved out to the other hill. The hill was mostly covered with thick thorny bushes that were pricking the hell out of everybody. I, as well as others, carried a machete (big, big long knives) that we would use to carve paths. Several of us moved up behind the point man and began hacking away.

After we had carved a reasonable path, we stopped for awhile, dropped our gear and proceeded to carve away a section in which to set up. We took turns chopping away at the bushes. Apparently, this was a virgin section of the hill. There were not any signs of passage through here or any other activity for that matter. That was okay by us. That meant that it would not be a possible traffic area for the enemy to pass through. We could easily check it out each morning before we moved back in to make sure it was not "booby trapped" the night before.

Once that was accomplished we sat back and lay around. This war could best be described as long periods of tedium and boredom permeated by moments of pure terror. We had heard that about other wars as well. We spent a lot of time waiting and watching. This time we had to stay back while the big artillery and the F4 pilots did their thing on the front side of Hill 474.

We played a lot of cards. Sometimes straight poker, but mostly we played a game called Tonk.

Tonk was like five-card gin rummy. You had to get a three-card spread, then a two-card spread, to actually "tonk". However, you could get a three-card spread and call using the value of the remaining two cards to establish your points. The stakes were three dollars for a "tonk", one dollar for a "call" and two dollars for a "burn". A "burn" happened when someone would call and someone else had fewer points. It was a rather simple game, but it was a lot of fun, easy to play and a great way to pass the time.

We had played a couple of poker games before, but not with any consistency. Tonk became a major part of our daily routine. Myself being an ardent gambler, I immediately jumped into these games. With the new guy carrying the extra radio, I passed the duties of monitoring the horn to him. The professor, who had also elected to keep his radio and remain the lieutenant's RTO, would take the duties once in awhile. For the first time, I was able to exercise rank. It was great not being the low man anymore. Hell, I was an old pro now.

During the day we received a shipment of packages from home that had been lost for over a month. Actually they were not lost, but were up north with the main element of the 101st. At any rate, they were a welcome blessing. Except for the fruit and cakes all was still intact. We immediately buried the fruit and cakes. Can you even imagine the stench of month old fruit and homemade cakes that had been sitting on a hot tarmac (heli pad) in the blazing sun? We were surprised they even sent it on.

The rest of the packages offered an assortment of canned goods that even rivaled the feast of Christmas Day. We immediately dug into our rucks and threw out all of our c-rats. We would have another feast back on top tonight. We usually moved back to the top around four o'clock, which was usually three hours before dark. It would be plenty of time to whip something up.

The package I received was from the girls at the office where I had worked back in the world. The package was a large box, loaded with all of the various canned goods you could imagine, including pudding in a can. There were packages of drink mix to "color our water, something other than brown", the pre-sweetened kind. There were also two cans of a pre-mixed cocktail drink "for those cold nights". Those I would save for later. It was like Christmas all over again.

It is hard to believe that a simple thing like a can of pork and beans can suddenly become a delicacy. I cooked those for lunch.

I had also received a camera from my mother, one of those little instamatic ones. I proceeded to take pictures of the guys in my platoon and a couple of the air strikes taking place on the other side of 474. I was able to catch the F4 fighter jet as it approached and again after it had dropped its bombs and was pulling back up. They were a couple of great action shots.

I had a couple pictures taken of me. However, one thing I forgot was that I had my hand bandaged from that incident in the truck. When I sent the pictures home, accompanied by a letter, I forgot to mention why my hand was bandaged. That caused a few problems later trying to explain. Oh well, you can't win them all.

Around 3:30 that day we started to pack up and prepared to move back up to the top of 474. I did not realize the added weight of the goodies from home until I tried to lift my ruck. I could not get up. The ruck usually sat on the ground in an upright position and all you had to do was slide your arms into the straps and stand up. Normally that wasn't too hard, even on re-supply days. A buddy would pull to help get you up or you could grab a tree or rock and right yourself. When that wasn't available you used a leg movement to stand up.

This time I could not get up. I wasn't alone, several others were having problems as well. We all eventually helped each other to our feet and I was finally standing, but the weight of my ruck was crushing.

Standing was the first problem. The next was getting back up that hill to the top of 474. If it was difficult to stand up, you can imagine that it was even more difficult to climb uphill. The trek was extremely slow going. The CO was at the top making sure everyone made it. As I approached the CO, he extended his hand to help me along with a friendly kick in the pants to encourage my continuing onward. I decided I would eat a very big meal that night to alleviate some of my load. God, was that ever heavy.

Once we moved into our places for the night, we began preparation for another feast. Mixing spaghetti-o, beef-a-roni and beans and franks into a steep pot (helmet), we ignited a fire under it and waited for the proper temperature to devour our delicacy.

After our feast, we laid around waiting for nightfall. We would have to pull guard shifts, but these would be a little easier than the ones we usually pulled in the boonies. Basically all we had to do was watch for anyone trying to come up the hill.

The top of 474 had been prepared for a large "kill zone" which is a large area cleared down each side to allow a clear view of anyone approaching. No one could sneak within fifty yards or the top without being spotted.

Guard duty was a rather easy task, especially with continuous illumination flares popping in the sky. They wanted to keep the hill lit up at night in order to help all ground forces see what was approaching.

As we were watching we spotted a firefight erupt on the bottom. The US and ARVN forces typically used red tracer bullets. The NVA and VC typically used green tracer bullets.

We spotted a barrage of red tracers fly across an open area at the base of 474. They were met with a barrage of green tracers. We continued to watch as the red tracers fanned out to form a U shape around the green tracers, then into a complete circle of red tracers. You would believe the ARVNs had the NVA/VC surrounded, but it looked more like they were shooting each other.

The next morning we found out they did wind up shooting each other. The ARVNs had come across a few enemy soldiers trying to make their way out of the hill. The enemy soldiers had been killed instantly, but the ARVNs weren't satisfied so they continued attacking until finally someone realized they were shooting each other. As I said before, we were glad they were down there.

Jack and I joined a couple of other guys who were passing a joint around. We had gone through a couple and were starting to feel even more relaxed. The explosion of the artillery shells that had started pounding the side of the hill was muffling the sounds of our conversation.

We were debating how easy this would be for awhile, even with our having to sweep down afterwards. With the amount of destruction they were putting on the hill, we doubted we would run into anything we couldn't handle.

Afterwards, Jack and I sat around and talked awhile. This time Jack was asking all the questions and I felt like the old pro.

"What do ya think about smoking in the boonies, kid?"

"Only when it's something like this, Jack."

"Huh, what do you mean?"

"You know … secured."

"You call this secured?"

"Well not really secured, but not as bad as being in the boonies, where ya have to be on your toes alla time."

"Hey kid, aren't ya scared you'll get caught?"

"Caught…? What caught?"

"You know, smoking dope."

"We haven't been caught yet."

"Yeah, but we've been in places where it didn't matter."

"Don't really care."

"Don't ya kid? Why not?"

"Just don't. Smokin' relaxes me. That's all that counts."

"Do you know what they can do to ya?"

"Don't give a fuck, Jack."

"Wow, man. They bust ya and then ya gotta do bad time. Then ya gotta spend more time in the army. Ever think about that?"

"No. I'll worry about it if it happens. Who really gives a shit if ya smoke or not?"

"Some lifer, who's a prick, looking to get points for himself. You know there's always one of them around. Doesn't that worry you? The sarge is probably one. Never touches the stuff. He's always straight. Maybe he should then he might not be such a prick alla time."

"Wow, Jack. I didn't know you hated the sarge that much."

"Don't, just his bullshit. How do you take it being his boy?"

"What do you mean, boy?"

"You carry his fuckin' radio don't ya?"

"Yeah, but that don't make me his boy. Shit, I never thought of it like that."

"Then, why do you do it?"

"I like the radio. It beats the machine gun."

"You coulda gave it to a cherry boy."

"I told ya, I like it. It's not too bad humping the radio. I'd rather be the lieutenant's RTO, then I wouldn't have to take as much shit from sarge."

"Don't you like the sergeant boy?"

"Com'on Jack, don't fuck with me like that. I don't like it. I ain't nobody's boy. I just like carrying the fuckin radio and that's all there is to it."

"What do you think of the lieutenant?"

"He's pretty cool for a louie. I think we can trust him."

"He's stupid."

"He's what? Jack how...?"

"I said he's stupid."

"He shouldn't let sarge push him around like he does, but I think he needed sarge at first, to help him get started, then it got out of hand."

"Regardless, he still should put the sarge in his place. It ain't cool with sarge around. He's a threat."

"Jack, are you...?"

"No. Shit kid, nothing like you're thinking. I don't wanna blow him away. He's not that bad. Just we gotta convince the louie to put the sarge in his place."

"Wow, you had me wondering. If he did that don't ya think the sarge would be out to get us then?"

"Nah, he's bucking for another stripe. He won't fuck with nobody till he gets it. Besides, he's not all that well liked. Somebody else might set him up and believe me he doesn't wanna jeopardize that promotion no way. The sarge will never get bad enough to blow away. Why ya nervous kid? Ya like him that much?"

"Hell no, but I don't dislike him enough to do him in or see him done in. Why all the questions man? You're starting to sound like me."

"How does it feel having the questions fired at you for a change? Bugs the shit outta ya doesn't it? Huh?"

"Yeah, as a matter of fact it does. I know what ya mean now, but why?"

"Ya wanna know the truth?"

"Yeah. Why?"

"Well, shit, when ya didn't give your radio up, with all them cherries around, I got to wondering what side you was on. I mean after all, no one wants an extra load. Christ, you couldn't even get up…"

"You noticed…"

"Yeah, I noticed. You was struggling when ya didn't have to. I got to thinking to myself. Why would he want the extra weight unless he wanted to or was stuck as the sergeant's boy…?"

"Jack…"

"Let me finish. With the sarge being as straight as he is and you asking me all those fuckin' questions I was beginning to wonder where you were at. You know what I mean? Here I was letting you know everything I know including smoking and showing ya sources. How was I to know you wasn't reporting back all that shit to the sarge, settin' me and the others up for something?"

"Jack, you actually think I would do that? How can you be so fuckin' stupid for Christ sakes? You really believe I would do that to ya after all we have been through? Boy, you think ya know your friends… shit… you really are fucked up this time… shit… and I thought… shit…"

"Gets ya, doesn't it kid? Remember what I told you back at the firebase about trusting someone? Well I trusted you, that's why I'm even talking to you now. I hadda let ya know what I was thinking so you would trust me. You know what I mean?

Look, I didn't have to tell ya nothing. I coulda kept quiet and let ya think nothing was wrong, but you and me's been pretty close and I wanted to let ya know what I really thought about all them questions. I tried to shut ya up before, but you kept pushing. I would have let it drop, but when you kept that fuckin radio I hadda make my move. You know what I mean kid?"

"Yeah. Were you thinking about blowing me away, too?"

"Nah. Shit. You wouldn't been worth it. Besides I told you, I like you. Look I said I trusted you, ain't that enough? Let me tell ya one thing else. Watch out for sarge. I don't think even you can trust him. As long as you're his boy, he'll let ya ride, but don't do nothing he could get ya for later. You know what I mean?"

"Yeah, don't worry. I won't let him get me, especially not now. Can I ask a question?"

"Sure kid."

"Aren't you scared of getting caught?"

"Not anymore."

With that we shook hands. Jack had made me angry at first, then scared, then angry again. I did trust him and I hoped he trusted me. My God, hadn't he said you have to trust someone? I wasn't a fink. I couldn't be, especially not against my best and closest friend. How could he think such a thing? There had to be a reason, besides my not giving up the radio. I had to ask.

"Jack, why else did you think that?"

"I told you, the radio."

"No Jack, there's something else, another reason. What is it?"

"The radio. Now get off my back."

"No man. I don't believe it. There's something else and I wanna know what the hell it is."

"Something I heard."

"What?"

"Something."

"What, com'on. You gotta tell me. You owe me that much."

"Something the sarge said to the CO."

"What?"

"Something about you taking care a it."

"Taking care of what for Christ sakes?"

"I dunno. Something you're gonna take care of."

"I still don't know what the fuck you're talking about."

"I heard the sarge tell the CO not to worry about that problem, the kid will take care of it for us. That's all I heard. What did he mean?"

"Hell, Jack. I don't know... I really don't know... The kid will take care of it... I really don't know... Was there anything else they were talking about?"

"Yeah, they said something about a list of names."

"Ah ha... you stupid fucker... you stupid mother fucker... Why don't you ask somebody before you jump to conclusions."

I pulled a list of names out of my pocket and began laughing, unable to explain what it meant. Jack looked at it questioningly, unable to ascertain what it meant. I knew the names on that list did not fit any pattern he might come up with, but I was still laughing too hard to tell him what they meant. Jack became caught up in my laughter and started laughing with me.

When the laughter finally subsided, I tried to explain, but every time I started to speak the hysteria of it all came back and I started laughing again. I could not get out more than a few garbled words before it would happen and Jack would start back up with me. He still did not know why he was laughing.

When I finally got out enough words for him to understand what the list of names was, he laughed even harder. We both were muffling our faces in our poncho liners to keep from making too much noise. When we had laughed until our sides hurt, we finally quieted down enough to speak.

"Admit it Jack, you really are a stupid mother fucker."

"I really am a stupid mother fucker."

"To think you would a blown me away for nothing."

"I wouldn't a kid, but you should blow me away for thinking what I did."

"No man, I wouldn't give you that satisfaction. I would a let ya go through life being a stupid motherfucker. It was comforting seeing you the fool for a change. You acted just like a goddamn cherry boy for Christ sakes."

"You are right kid. You mean that's all that list is?"

"Yep, just a list of those who didn't get a write up for a medal. Sarge checked the write ups and made a list of the names of those who didn't have one, so he told the CO he'd get them and naturally me being his boy, as you pointed out, I got stuck doing it. And you thought… Boy you really are…"

"Yeah I know."

"Well you are."

"You gonna let me live this down?"

"No."

"I didn't think so."

"Think I'll spread it around."

"Com'on kid. What ya wanna do that for? It's bad enough you knowin without everyone else… com'on gimme a break."

"Alright, Jack, just remember, before you go jumpin to conclusions again shithead."

"Yeah… yeah… don't rub it in."

Jack had actually jumped to conclusions over some piecemeal information he had overheard. I sympathized more than I blamed him. He was really a close friend after all for having the guts to even talk to me about it. We were continually binding our friendship. We had arrived in the country together, which meant we would hopefully stay together for the whole tour of duty. We would not be so lucky.

The next day was the same, except for the packages. Around 3:30 we moved back to the top. It was easier this time. I had eaten some of the food. My ruck was not as heavy. I even avoided the CO's kick in the ass.

That night after we ate, the lieutenant picked Jack and a couple of others to set up an OP (observation post) outside the perimeter just a few yards away. They were close enough to be in speaking distance to us, but Jack wanted a radio anyway. He would pull the first shift, then come back in for the night. It was easy enough.

The shelling of the side of 474 had been shifted more toward the top and we were getting small pieces of shrapnel flying over and around us. The lieutenant had me call the CO and suggest they lower the target. Our area of the top seemed to be getting the most of it. The CO was already on it.

I sat there listening to the traffic on the horn. After it quieted down I heard Jack's voice come on.

"Hey kid, you there?"

"Yeah, com'on man, ya gotta use the call signs."

"Fuck the call signs, I'm hit."

"You're what?"

"I'm hit. I caught a piece of shrapnel in the wrist. Would ya send the medic over? It hurts like a mother fucker."

"Yeah man, hang in there. I'll take care of you."

I got the lieutenant and the medic. I explained what happened. The lieutenant said he was going over there with the medic and another guy to take Jack's place. They would bring Jack back over here. I called Jack back, informing him to expect the LT and the medic and to hang in there. I waited.

In a few minutes the lieutenant called back and informed me the medic thought Jack should be sent in right away. I called the CO. He had heard. A dust off was on the way.

On the way back, the lieutenant fell in a foxhole and broke his wrist. He was going in also.

Jack came over and sat by me. We looked at each other for awhile. I stared at his wrist. There was a rather large (large in comparison to his wrist) piece of shrapnel lodged in his right wrist. It was bleeding quite heavily.

"Well ya really did it this time, Jack. Now what are you gonna do?"

"Go back to the world kid."

"Then what?"

"Whatever... I might be back."

"Don't press your luck."

We could see the spotlight of the dust off approaching. Jack would leave soon. We shook hands and stared at each other again.

"Jack, can I ask one last question?"

"Yeah, sure kid, go ahead."

"How old are you?"

"How old am I... why?"

"Just curious. You're the one who tagged me with kid and I wanna know how much of a kid I am."

"You're gonna blame me for that too?"

"Nah... how old?"

"Twenty-two."

"Twenty-two?"

"Yep, just turned."

"You son-of-a-bitch and you call me kid?"

"Yeah, why, how old are you?"

"Twenty."

"See you are a kid."

We would not be able to finish the conversation. The dust off was making its approach and we had to grab everything to keep it from blowing away. The helicopter landed and I watched Jack and the lieutenant scamper aboard.

I watched the helicopter take off and I continued to watch as it flew off into the night.

There went two more guys. That made nine of the original twenty-five we started with lost to this hill. There were only sixteen of us left to complete our vow.

As I sat there thinking about losing Jack and the lieutenant, sarge came over and spoke to me.

"What's up kid, miss Jack already?"

"Nah… just thinking, that's all."

"Want to talk about it?"

"Not to you."

"What's the matter, don't you like me?"

"No, as a matter of fact I don't."

"Why not? I ain't never done nothing to you."

"Not yet."

"Com'on kid, we can be friends, just like you and Jack were."

"No sarge, I don't think we can. Now would you leave me alone?"

"Sure kid. Just let me know if you need anything."

It was hard to tell if the sarge was sincere or not, but I could not take the chance. Jack had warned me about him and it seemed strange that he would suddenly want to be my friend now that Jack was gone. Ah fuck Jack, why did you have to leave? Just when things were looking up, you go and split. It ain't fair. It just ain't fair. My thoughts were heavy.

We would get a new lieutenant tomorrow. I would have to tell him about the sergeant. I owed Jack that much. He had warned me and I would warn the lieutenant.

Sleep would not come easy. I had lost the only one I trusted over here and it would be hard to start over. I was no longer as naïve as I was upon arrival and I would not take to anyone so easily.

The following morning we stayed on the top a bit longer than usual. We had to wait for the arrival of the new lieutenant. They didn't waste any time. Sarge was in his glory, however brief it may have been, because he was in charge until the lieutenant arrived. I tried as best as possible to avoid him.

We did not have to wait long. The new lieutenant arrived about 10:30 am. Along with him were a couple of new guys. One of them, after exiting the helicopter, stood there bewildered as if he had forgotten something.

There had been supplies aboard that had to be unloaded and the two new guys had helped. Apparently during the shuffling the new guy had assumed his gear was pulled off. It was not. Not only was he missing his gear, but also his rifle. We had to bring the helicopter back. Needless to say, the pilot was not happy.

One consolation, I still had the rucks, which were formerly used by Jack and the old lieutenant. I might as well send them back in, I thought, instead of field stripping them.

As I waited for the helicopter to come back, I remembered to look in Jack's ruck. His "stuff" was still there. I pocketed it and gave the ruck to another guy to load onto the helicopter. I had to monitor the horn.

## THE PROTECTED WILL NEVER KNOW

I rubbed my pocket as I sat there and made a mental note to have one for Jack. I was sure he wouldn't mind my taking his stuff. In fact he probably counted on it, so that it wouldn't be found in his ruck later. In any event, I would take care of it.

After all the problems were taken care of we packed up to move down off the top. As soon as we were set up the CO brought the new lieutenant around and introduced him to everyone. Of course, we regarded him with apprehension. No matter what, he was still an officer and we had to be careful until we found out where he stood.

It had been known that some of these "first time in the boonies lieutenants" weren't too wise. They had been known to make some classic blunders. Even though they were officers and we had to follow their orders, there were ways of straightening them out. The new LT appeared all right, but time would tell.

I noticed the sarge was taking advantage of the situation and was trying to put the new LT under his wing immediately. Somehow I had to tip the new LT.

After things had settled down, the LT called a meeting of the sergeant and the squad leaders to discuss our plans to sweep down the hill. I noticed the sergeant monopolizing the conversation and trying to sway the LT toward his way of thinking. This lieutenant wasn't giving in as easily as the other one had. Maybe he was on to the sergeant already. I still had to warn the LT.

Later on in the day, I had a chance to talk to the lieutenant. I didn't want to tell him right away, but somehow I had to drop a few hints.

"So you're the sarge's RTO?"

"Yes sir, I am."

"How do you like it?"

"What do you mean sir, sarge or the radio?"
"Either?"
"I like carrying the radio."
"What about sarge?"
"What about him?"
"What kind of guy is he?"
"I'm not sure what you mean, sir."
"Besides being a prick. What else is wrong with him?"
"He's straight... I mean..."
"I know what you mean. He doesn't smoke."
"That's not exactly..."
"Hey, cool it. He's probably the only one who doesn't. What's your pleasure?"
"Ahh... ahh..."
"Com'on you can trust me."
"Sir, I think I better see if the sarge needs anything."
"He doesn't, besides, I'm in command here. What do they call you?"
"Kid."
"Why?"
"I don't know. Everybody started calling me that after Jack tagged me with it."
"Jack...? Oh yeah, I saw him in the hospital this morning before I came out here. I saw the lieutenant too. He warned me about the sergeant. So did Jack. Jack said I should talk to the 'kid" about the sergeant, that you would give me the lowdown. Will that clear me with you?"
"Shit... I mean... Wow sir, I don't know what to say."
"Oh kid, how'd you like to be my RTO?"
"I'd love to, but that wouldn't be fair to the professor. He's the senior RTO and deserves the easier... ah... working with you."

"The easier job huh? You think it will be easy working for me?"

"Well, you know, sir. It's better than the sergeant and I couldn't do that to the professor."

"Let me work on it, see if I can make it better for everybody. Nice meeting you, kid."

"Yeah sir, same here, catch ya later."

Wow, even when Jack wasn't here he was still taking care of things. I wonder why he trusted this lieutenant so easily? Well, the last one was okay, maybe he just had one of his feelings

Speaking of feelings, I wonder if he had percentages on himself. After all, it was his game. Hell, he probably knew all along when he would get out of here. That's probably why he took me through that ordeal the night before, to set things straight before he left. Damn him anyway. Thinks I can't get along without him. Well maybe I can't, but I have to try sometime.

The new lieutenant seemed all right, probably because he met Jack and Jack set him straight about the sergeant and me.

We spent a few days going up and down. Finally, the day arrived to start our sweep down. During this time, the professor and I were sharing duties for the lieutenant, which meant the new guy with the reserve radio kept getting stuck with the sergeant.

It only seemed fair, except to the sarge. Sarge was beginning to realize what was happening, although he did mellow somewhat. The sarge did not try to continue to be a one-man show. The lieutenant even joined in some of our card games. He was a gambler like the rest of us.

It had been a fun week. Mostly just playing cards and generally cutting up. Tomorrow we would have to go back to work. That was all right, we had a nice break in the action. There was some hassle over who would go with the sarge. The professor, lieutenant and myself decided it would be best for everybody if I did.

We would be leaving first thing in the morning so everybody got to sleep early. We listened to the last of the artillery shells pounding the side of hill 474. It was like being rocked to sleep.

# SWEEPING DOWN

**WE STARTED DOWN SLOWLY.** We were still apprehensive about running into something, or someone.

We had barely gone fifty yards when we hit another complex of rocks very similar to those we were in that fateful day. The only real difference this time was the lack of vegetation. The area was entirely clear of any jungle growth.

There were obvious cave entrances that had to be checked out. Just to make sure, a hand grenade was lobbed into each one before entering. Now everyone wanted to be a tunnel rat, including myself.

Our platoon had set up in a little ravine. It was the most level spot around and we were sending groups out so far to check every inch of ground. We had split the platoon into three groups, so that each group had a radio and RTO with them. We left one behind and the other two branched out to the right and left. The professor was staying back this time with the lieutenant as we had agreed and I went with the group, which branched right, with the sergeant.

## THE PROTECTED WILL NEVER KNOW

We had ordered extra cases of grenades and each guy was heavily loaded down with them. We had left our rucks on top, as we would be going back there every night. It would be easier to move up and down without it. Besides it would be near impossible to set up for the night on the devastated side. Even if we went more than halfway down, there would still be enough time to get back up. The hill was approximately 474 meters high, hence the name Hill 474.

The lieutenant was carrying a forty-five pistol as well as his rifle, but was not using it. He had it packed away in his ruck. I asked if I could use it. I explained that with carrying the radio, it was awkward to use my M16 and the radio while searching through caves. It would be useful to anyone in the group I was with and we could share it. He agreed and gave me the forty-five. The lieutenant even had a holster with it.

I fastened the holster to my belt and used a piece of shoelace to tie it to my leg. I had to fasten the holster to my right side even though I was a left-handed shooter. That was all right though since I had qualified on the forty-five both right and left handed during training. So it really didn't matter which hand I had it in. If anybody else wanted to use it, they would give me their rifle, so I wouldn't be defenseless. I really liked having that pistol.

While I had the radio on my back, I would always keep the handset close to my right hand enabling me to answer right away. This caused a problem with handling my M16. Even though the M16 could be fired effectively one handed, it still left me in an awkward position, trying to do both. With the forty-five pistol, it would allow movement of both hands.

Just for personal satisfaction, I practiced drawing. The forty-five is not the best pistol to use in a gunfight, but I had developed a rather quick move that extracted and had it in my left hand in a matter of seconds. If necessary I could use my right hand to fire, but it felt more comfortable to use my left.

I really enjoyed carrying the forty-five. I would only take it off at night. The lieutenant apparently didn't mind because he never asked for it back. The rest of the guys enjoyed having it too. The forty-five made it easier to search caves.

On the second day of searching, I stayed back with the lieutenant.

Our Bravo (B) Company was approaching the area in which we made contact. They were to get there first, so that they could remove the bodies before we got back there. Although we knew each other from platoon to platoon, we really didn't know the guys in the other companies. It had been decided it would be better for B Company to pick up the bodies we had left there.

B Company called up to clarify the location and the lieutenant passed the horn to me. The new lieutenant had not been there. I tried to explain what it looked like then, but I wouldn't know what it looked like now after the side of the hill had been devastated from all of the shelling.

As we were talking, B Company started taking fire, not from rifles, but from an M79 grenade launcher. The M79 was similar to a rifle, but it fired grenades. At first they thought it was one of our own units, because they knew the gooks didn't have that type of weapon, but I explained that we had lost one that day we made contact. B Company decided to wait awhile and called a gunship out to quiet things down before they moved in.

Once the gunship finished, B Company started back in, but could not find the bodies. They estimated they were in the right spot, but there was no sight of any of them. We had left the bodies scattered on the path and B Company should have found at least one by now. B Company said they found a lot of spent rounds, shell casings and a shot up handset from a radio, but nothing else. B Company had to be in the right spot. They said they would keep looking and let me know if they found anything.

I went back to the business at hand. Each of the two elements reported in. Neither had found anything yet. I knew they were listening to the conversation with Bravo Company. They would want to know also and weren't really looking.

Bravo Company called back, they had found the rocks, but still no bodies. Bravo believed they also found the path we were talking about and could we give them any more information. I suggested they keep searching the area around the rocks, maybe the bodies were moved for some reason. Bravo agreed and said once again they would get back to us.

I heard the CO come on the horn. He suggested the same thing. The CO explained that the NVA had a high respect for the dead and would take all precautions to preserve the bodies from further mutilation. He also said that if the VC gets a hold of the bodies we might never find them. The VC had a reputation of mutilating the dead. They would do grotesque things to let us know we had met up with VC and not the NVA, but to keep searching anyway, the bodies had to be there somewhere.

About a half hour later, Bravo called back. They had found the bodies. The bodies were wrapped in ponchos and tucked into a rock formation for protection. Bravo had unwrapped one of the bodies and said it looked okay. The bodies had not been mutilated. Bravo also said they had ordered body bags and were going to send the bodies in. Afterwards Bravo would pull back to their position.

I thanked them for their persistence and promised Bravo if they ever needed anything to give us a call. I watched the lieutenant nod his head in agreement. Bravo had taken care of our vow for us. The bodies of Jeff, Benny, Tom and Guam had been recovered and sent back.

We watched the helicopter come in and drop off the body bags. Then the helicopter hovered while Bravo Company loaded the bodies into the bags. In a few minutes the helicopter went back to the spot. We could barely see the top of it while Bravo loaded the body bags on. Shortly the helicopter took off again and Bravo came on the air saying that everything was taken care of. Bravo would be moving back into position. Glad they could help. I watched the helicopter disappear in the distance. I said my last goodbye to the four we lost that day. They would finally be sent to a proper resting place.

After that, everyone was able to go back to what they were doing, even the lieutenant understood and did not press to continue searching. One of the guys in our platoon was a cousin to one of the dead, so he was called to be a body escort back to the states. He would have to come back, but it was a chance to get out of here for awhile. We had not advanced very far down, so he went back up to the top by himself.

I had been fondling my "drive on" rag that hung from my neck during the search for the bodies and the lieutenant asked me what it was for. I explained to him why we wore it and what it meant. I also told the lieutenant that there were only sixteen of us left. He said he had noticed that only half of the platoon wore it and that is why he had asked.

Among the various odds and ends we had collected, mostly junk, was a ripped poncho liner. The lieutenant picked it up, ripped apiece off and tied it around his neck. As I watched him, he explained that he wanted to be one of us. I said I thought it would be okay and we shook hands on it. Some of the other guys that were there with us agreed and also slapped hands with the lieutenant. Actually the lieutenant was one of us and we felt that he should be part of us as well.

The next day, it was my turn to go out and let the professor stay back. The new guy did not have a choice in the matter, he had to go out every day. Privileges of rank.

We moved further down the slope and encountered another cave complex. After dropping several grenades into the structure, our various tunnel rats entered. The rats came out seconds later beaming with joy. They had hit pay dirt.

The rats found ammunition, rifles, grenades, a couple of rockets and launchers and even a typewriter. This was our biggest haul yet.

As I was relaying the information back to the lieutenant, the CO came on the horn and asked if it looked like anything else was there. I responded I didn't know, but the rats were still down in the caves.

Just as I finished speaking another guy emerged saying he had found a lot of documents, but they were all in Vietnamese. The CO replied that he would send some guys down from the second platoon to bring the documents up, so he could get the documents back to intelligence right away. The CO reiterated we should bring all of the documents up out of the cave first. The rest of the stuff we could do with what we wanted.

There were some personal effects, such as NVA money, which I confiscated and put into my pockets. There was also some NVA equipment. Everyone was grabbing something for their souvenirs. We had made a pretty good discovery. This had been the best find so far.

We continued finding odds and ends and I kept relaying back to the lieutenant what we had found. We had a signal arranged between us in case we found anything he wanted, but so far nothing peaked his interest.

The guys from the second platoon made it down and each guy grabbed a handful of documents to carry back up to the CO's position. There certainly were a lot of documents. They looked like files of some sort. Each was a separate folder that contained various pieces of paper. Oh well, let intelligence figure it all out. That was their job.

The CO called and suggested we send the rifles, rockets and rocket launchers back with the guys from the second also to be sent in. We could do what we wanted with the ammunition. I asked about the typewriter and the CO asked me if I'd like to hump it with me. When I said an emphatic no, he said either send it back or destroy it.

I first decided to destroy the typewriter, but then changed my mind and gave it to one of the guys from the second to bring back. The guy said he would rather I destroy it, but I insisted he take it. I told him the CO called back and said to send it in. I didn't know, maybe intelligence could get something out of it. It isn't too often you find a typewriter in the boonies. This must have been some kind of headquarters for the enemy.

The next day we were going into the area in which we had made contact. Now that the bodies were removed we were allowed to go in there. We approached the area from a different angle and with the growth missing it was unrecognizable to us.

The way we approached the area took us through the LZ first and then back into the rocks. Off to the side of the path, I noticed a helmet lying against a rock. I picked it up. The inside was filled with dried blood. It was dark red, bordering on brown, mixed with army green. On the backside were the printed letters:

## THE PROTECTED WILL NEVER KNOW

Jeff. I handed the helmet to one of the guys who looked at it then passed it on.

There were other odds and ends on the path, such as bandages, a lot of shell casings, a shirt and pants and the broken handset to name a few. The path looked a lot shorter now. It had seemed extremely long that day.

The rocks even looked different. At first we wondered if this was the right place, but we knew it was. The escape route the lieutenant and I had taken that day was now wide open. The path no longer had that covering we used to protect our getaway. The rock formations were a bit different. The bombings and shelling had left their mark.

The overall structure had been altered somewhat, but the cave entrances were virtually intact. Damn, there may still be someone in here. We glanced up the hill to the approximate spot we were taking the shooting from. It was ironic, the group we left behind was in the exact spot. We waved to them. They waved back. At least that was safe.

We started sending guys down into the caves. They came back out immediately. They had found our rucks. I went down with the second phase, but instantly developed claustrophobia and came back out quickly. This was not my game. I waited while the others went down and started bringing up our gear as well as the gook gear. We noticed our letters had been removed from their envelopes and mixed with others. The gooks obviously went through them. We recovered the two radios we lost and decided to keep an extra one with the platoon.

One guy came out and said he found a second set of caves that branched even lower. He wanted to take another guy and explore further. The lieutenant, who had come with us this time, agreed and we waited for their return. The tunnel rat had borrowed my, ah the lieutenant's, forty-five and proceeded back down.

When the rat finally emerged he was beaming with excitement. He had found a hospital complex down there.

There were intravenous feeding bottles and large pools of blood. No bodies, but apparently a lot of casualties had passed through. There was a large section off to one side that had lanterns and supplies of bandages. The gooks must have heard us coming and pulled out recently because some of the blood was still wet.

The guys went back down and searched some more. When they came back up, they said they found syringes and other hospital type supplies. We rigged it for demolition and waited until we were ready to pull out.

Before we left we rounded up all the gear we could carry and prepared to move out. We gave it one last look and moved down to what was the LZ. Upon looking at it again, we fully realized what those dust-off pilots had to go through to get in here and the target they presented to the enemy up the hill.

As we discovered back in the rear, the first dust off that came under fire had the crew chief seriously wounded and their helicopter was written off as a loss, it had been hit with so many rounds. When the second dust off hovered by the ledge for the pickup the enemy riddled it with bullets as well. No one was injured, but the bird was so damaged it had to land in the valley and could not make its way back to LZ English. The crew of the first dust off attempting a second pickup manned the third dust off. It too was hit by enemy gunfire and the pilot was hit, causing the bird to spin out of control, but the aircraft commander was able to regain the controls and land the damaged bird in the valley.

## THE PROTECTED WILL NEVER KNOW

Finally the next morning, a fourth helicopter attempted the pickup and was able to get the wounded men on board before pulling away, all the time being sprayed with enemy gunfire. The bird made it back to LZ English and the wounded men were hustled into the aid station. Meanwhile, the helicopter was declared another combat loss.

Standing on the ledge now with the foliage gone we realized what those pilots had to go through to get the helicopter close enough for a pickup on the ledge and how vulnerable they were to attack. As I said before, dust off pilots were the best there were in Vietnam bar none. A couple of us stood at attention facing toward LZ English and offered a silent salute.

When we got far enough away, we spread the word that we were going to blow the cave and proceeded to detonate the charges. The explosion started with a low roaring boom and culminated in a loud roaring blast. Once the dust settled, three guys went back and set some more charges to repeat the process. The lieutenant didn't argue with the second set, even though it probably was a wasted effort. He knew why we had to do it.

We carried the gear back up to the rest of the platoon and those original sixteen took their letters and destroyed the rest. We had all the reminders we wanted of that day. We did not need any of the old gear. Most of the old gear had been mixed up. The gooks must have tried to break it down and use it for themselves. We destroyed the extra radio, kept the second and listed them both as a combat loss. The thought was it might be advantageous to have four radios.

We looked at the rocks again and a couple of guys wanted to go back down and set some more charges, but the lieutenant vetoed that idea. He said we had done enough. Besides it was getting late, we should start making our way back to the top.

Our Bravo Company had been working the area below us and we had accomplished a fairly good sweep. Tomorrow we would work further to the left, staying parallel with the area we had covered.

The next morning, after we had started down, we were called back to the top. Our platoon would be broken into four groups, three of which would be sent to various points at the bottom to pull ambushes. The forth, including the lieutenant, would remain with the second platoon and the CO.

We would be brought down by helicopter in the afternoon, with plenty of time to get set up for the night. It was decided we had spent enough time on the hill, especially since we got back to the contact area. They had pulled the ARVN unit out from the bottom and put them on the other side of Hill 474. The ARVN's had not accomplished anything since that first night. OPs (Observation posts) had reported seeing enemy movement toward the village every night and it was time we stopped it.

The plan was for us to set up in three different locations to try to halt the gooks' nightly movement, or at least upset it.

We got our gear together and waited for the helicopters. The lieutenant and the sergeant were not going with us. We would be three groups of six with eight staying behind. I had been elected to be in one of the groups, as the professor got to stay behind with the lieutenant. I didn't mind (not that it would have mattered), I had never pulled an ambush before. It would be different.

Everyone kept referring to our mission as "pulling a 'bush". Some of the guys had done it before.

Before we left, the CO came over to wish us well. He also told us what we had run into on (in) Hill 474. He said we walked into the 8th of the 22nd NVA (eighth battalion, twenty second regiment of the North Vietnamese Army). Apparently we had

## | 156 | THE PROTECTED WILL NEVER KNOW

walked into their headquarters that day and were too close to just let go by.

By the end of February we would be successful in the 8th battalion's defeat, capturing men, supplies, weapons and a slew of valuable documents before finally moving on.

# PULLING A 'BUSH

**The helicopters finally arrived.** We climbed aboard and once again left the top, this time heading for the bottom.

Our platoon had been divided into four squads, with three actually going out on an ambush. Each group boarded a separate helicopter. One group would go to the far side, the other group would go to the near side and our group would go to the middle of the hill. This would enable us to effectively watch the entire bottom. Of course, it was possible for the gooks to slip through, but it appeared that we had good coverage of the area.

Once we landed, we started looking for a place to set up. As we moved into what we thought was a good position, both as a means of gaining a good vantage point and a place we could adequately defend, we realized we were extremely close to the squad on the near side. We decided, since it was getting late, it would be more advantageous if we joined forces for the night.

We met up with the other squad on a small ravine that was separated by a deep valley. There was a slight incline that rose up one side, but it was wide open and presented a clear view of anyone trying to approach us. By joining the two groups, we were able to set up a larger perimeter.

We got into position and dug shallow foxholes to give us an even better chance to defend ourselves. We mostly sat around trying to get in that last cigarette before dark. Although we did occasionally smoke at night, it was advisable not to do so. A lit cigarette acted as a marker of your position in the dark, even more so while you were lighting up. The initial blast of light from striking a match or lighter would appear to have the effect of turning on a neon sign saying: "here we are".

One of the pieces of equipment we had with us was a starlite scope. The scope looked like a larger telescope, but with an infrared lens, which enabled the viewer to see in the dark. One of the new cherry boys was looking through it, amazed at its effectiveness.

The cherry had been watching for some time when suddenly he straightened up and exclaimed that he saw something. The rest of us, of course, ignored him. He couldn't possibly have seen anything in the area he was looking. He had been focusing on the far side of the valley and that area was an extremely rocky section at the base of the hill. We had viewed that spot quite vigorously before setting up. Besides, he was a cherry and it was his first time in the boonies. However, he persisted.

"Hey, I'm telling ya, I see one moving in the rocks…"

"Com'on man, are you sure?" Our squad leader asked.

"Yes… holy shit there's two now… three…wait, yeah, three… I swear there's three of them now."

"Yeah, and what are they doing?", the squad leader asked as he rolled his eyes and shook his head.

"Just standing there… wait, they're looking back… hey there's another… there's four of them… they're starting to move…"

"Gimme that Goddamn thing." The squad leader jerked the starlite scope out of the cherry's hands.

In the chain of command, each platoon had four squads, or at least three and each of these squads had one man assigned to be squad leader. Rags was our squad leader. He had been in Vietnam for a little over six months, but with another unit of the 101st. Rags had been in the A Shau Valley cleanup in late '69 so he was a pretty hardened grunt by now.

Rags grabbed the starlite scope from the new guy and tried to focus it on the area the new guy was referring to. I watched as Rags adjusted the scope to his sight and closed one eye for a better view.

Rags was scanning the area when he stopped abruptly and returned the scope to the spot he had just scanned. He moved the scope frantically while trying to pinpoint the exact spot. Rags stared hard trying to actually make out what he was seeing. His closed eye burst open, then shut again as he leaned forward, trying to get a better view.

"There are four of them... they're just standing there... like they're waiting for somebody... quick kid, call the CO... see if we can blow them away... hurry..."

"Okay Rags, I'm on it."

I tried repeatedly to raise the CO. I couldn't even raise his RTO.

"No one's answering."

"Christ, four fucking gooks in the open and they're sleeping up there. What the fuck is this war coming to?"

The radio crackled, two people were trying to talk, finally the static cleared and the third platoon's lieutenant came on the air. I answered roger and tapped Rags on the shoulder to get his attention.

"Hey Rags, the third platoon's lieutenant says go ahead."

"Yeah, what did he say about the CO?"

"He said, 'tell him in the morning'…"

"Out fuckin' standing. Alright… put some M79 fire out there, then open that machine gun on the area. Something's got to hit the gooks. Now! …What the fuck ya waiting for?"

The guy carrying the M79 grenade launcher started firing rounds into the general area, one right after the other. The new guy, who also was stuck with the machine gun, lay on the ground and braced the gun to his shoulder, ready to fire. Rags hit him on the head yelling. "Now!" The new guy fired a quick burst, regained his hold and fired another. That instantly jammed the gun. The new guy struggled unsuccessfully to clear the gun, but it was hopeless. He looked up at Rags, who was now swearing angrily at him and the gun.

"Christ man, you fire short bursts, not the whole belt at once. Those fuckin' guns. I ain't never seen one work right. Com'on, can't ya fix the fuckin' thing? What the fuck did you stop for…? Fire some more grenades. Kid call somebody, tell them to stick this fuckin' gun up their ass. Com'on fix the fuckin' thing…"

"I can't, there's three rounds stuck in the chamber."

The new guy was frantically trying to clear the chamber. Rags snatched the M16 rifle from my hands and handed it to the new guy.

"For Christ sakes… Here use the kid's rifle, use something, just shoot the motherfuckers. Kid what're they saying up there?"

"The CO wanted to know what the hell's going on…"

'Tell him we're taking target practice. What the fuck does he think is going on?"

"Ah… I explained the situation to him, about the gun jamming and he said throw rocks if we have to, but don't let those fuckers outta this hill."

"Tell the CO, I think we got a couple. I saw two drop before the gun jammed... Hey keep putting' grenades in there... They haven't got up yet. The other two went behind the rocks. Shit, just tell him we're doing what we can with fucked up equipment."

I was relaying the information back to the CO's RTO when the trip flare on the high ground popped. I as well as everyone else, looked over immediately. We saw two figures standing in the illumination. We had placed an extra set of trip flares further up and those had been the one hit.

The grenades appeared to be short of their target, but a couple of guys opened up with their M16 rifles and I saw the two figures drop. The high area was actually by the other squad, so Rags let them handle the situation.

I continued relaying the information back to the CO's RTO. Meanwhile I focused my attention back to Rags and our dilemma. Rags was scanning the area with the scope again.

"Shit, I lost them. I can't see any movement. Kid, tell the CO I saw two go down and we must of hit the others. We had to kill something with all that M79 shelling. What's happening over there?"

Rags pointed to the high ground, dropping the starlite scope to his lap. The M16 rifle fire was still pretty heavy. Rags grabbed me by the shoulder turning me in the direction of the high ground. One of the guys from our squad was standing and firing his rifle at the high ground.

"Hey, you stupid fucker. Get down. What the fuck is he doing standing up? Call over there and tell that idiot he makes a good target."

No one could hear over the gunfire, so someone threw a rock and hit the man in the back of the leg. When he turned everyone motioned for him to get down.

The radio crackled and I listened while Rags pointed to the guy to get back in position. Rags hit the new guy on the head again asking if now that it was quiet could he fix the machine gun. The new guy shook his head no.

"The CO says he wants a full report in the morning and with all the shit we put down here, we better have something or he wants to know why."

"What the fuck did he think we were doing down here? Tell him... tell him we had... we had... ah fuck it... we'll go look tomorrow. They got anything over there yet?"

"No. They said the two went down, but they're not going out there tonight. If the bodies are still there in the morning, they'll let ya know."

"Yeah, okay. Do something with that fuckin gun."

"What?"

"How the fuck do I know. Fix it, kick it, something, but ya can't keep the kid's rifle all night."

"Hey Rags, I still got the forty-five."

"It's up to you kid, but I'd rather have a rifle."

"So would I, but it will do 'till the morning."

"Alright, everyone get some sleep. I'll pull first watch. Sign off kid and gimme the horn."

I finished and let them know we were starting our watch. I handed the radio to Rags. The whole ordeal had taken about a half hour. The answer would come in the morning when we looked for the bodies.

The enemy was extremely conscious about our finding their bodies. They would move the bodies, hide them and they would get rid of any evidence that we had killed or even wounded anyone.

The Army was obsessed with body counts, the more the better. In fact, sometimes we would dig up graves to find out who was buried and how that person died. One time while we were searching the hill, we came across an area of fresh ground that looked as if something was buried there.

We were ordered to dig it up and find out. When we uncovered the ground, we did in fact find six bodies buried. One strange thing about two of the bodies, on average the Vietnamese man stood five foot to five foot two inches in height and weighed about one hundred pounds. These two bodies were at least six feet and weighed close to two hundred pounds. One of the bodies was wrapped in silk cloth, almost as if it was some sort of ceremonial burial.

Someone said it was the way the gooks buried officers, Chinese officers. The body could have been Chinese, but we didn't know for sure. To be thorough, we stripped the body and reburied it. Let intelligence figure it out. Believe me, it was bad enough just digging the bodies up. The stench from week old unprepared bodies convinced us that we should finish the job quickly and not worry about details.

At least some gooks didn't get out of the hill. Those were the first enemy bodies we had found while sweeping down the hill. Before the week was out, a lot more were found, buried or otherwise.

The next morning we had breakfast first and then prepared to start our search.

The area where the trip flare had popped was searched right away, mainly because it was so close. You could almost tell there wasn't anything up there, but upon searching the guys found quite a bit of blood and drag marks. I informed the CO, but he wasn't impressed. The CO explained that we needed something to send back in.

We went to search the other spot, hoping we would be more successful. Only three of us went. Rags, the new guy with the rifle and myself for radio contact. The spot didn't look that far away and we would be in constant sight of the others.

We worked our way down and across the valley. As we started the climb up the other side, we noticed we were getting quite a distance away. If we should hit trouble, we would be too far away for help. Rags decided to push on a little further. When we were as high as the hill we left and still had not found the area, Rags decided to go back and try later.

We backtracked through the valley back up to our secure spot at the base of the hill. I informed the CO of our progress. The CO was beginning to wonder if we had really seen anything.

In the afternoon we went back. This time there were six of us. We got back to the spot where we were at in the morning and started making our way across the side. We had a guy posted back on the ravine to guide us this time.

It wasn't long before we found a path. The path led us right into the spot. Everyone had stopped and were milling about. As I entered the area, I saw what everyone was staring at. There it was, a gook body. I called the lieutenant and informed him that we had found one. We were not seeing things after all.

The body was in a crouched position. The gook had apparently gone behind some rocks after we opened fire. He must have tried to look over the rocks and been hit by an M79 shell. There was a gaping red hole where his forehead used to be.

On the other side of the path, there was a large rock with a lot of bullet holes in it. The machine gun had been very effective against the rock. There was quite a bit of blood on the rocks and in the grass. We must have hit more than one, but this would do for now. As I stood there, Rags came over.

"Okay kid, go ahead."

"Go ahead what?"

"Strip it."

"Strip it?"

"Yeah, strip the body, so we can send the shit in."

"Why me?"

"Why not, you're the newest one here."

"No, I'm not... what about...?"

As I said it, I looked around. I was the newest. The cherry boy had not come with us. Rags went with a seasoned group. My eyes appealed to Rags. I stared at him. I didn't want to touch it. Rags read me.

"Sorry kid, but we've all done it before. Here, I'll take the horn."

"You're all heart. How do I get that rifle out of his arms?"

"Break it."

"The rifle?"

"No asshole, his arms. The rifle makes a good souvenir. He doesn't need his arms anymore. Hurry up so we can get out of here."

Rags leaned back against a rock and set the radio beside him. He made a motioning movement with his hands encouraging me to get going. I approached the body. The maggots were already having a field day. They had infested the wound and were looking for new avenues of entry.

I first tried prying the rifle loose, but he had been dead for a long time and the body was rigid. I grabbed the rifle with both hands, placing my foot against his chest and pulled. I heard something snap and one side came loose. The rifle was still wedged in his right hand, but the one side was free, so I kept twisting until the hand would not bend anymore. Finally the hand gave and the rifle was free of the body.

This was the first time I had seen an AK47 up close. It was a Russian made, Chinese copied rifle, similar to our M14, in that it had a wooden stock. The rifle was also bored so that they could use our ammo, but we could not use theirs. I checked the side. The side facing the body was covered in dried blood. I wiped it against the body's shirt and checked again. This one was Chinese or at least had Chinese markings on it. I passed the rifle over to one of the guys who slung it over his shoulder.

The body was dressed in full gear and I started pulling things off and throwing them onto the path. I would take inventory later. I ripped the shirt and pants off, although I did not take his undershorts off. I did not see any need for intelligence having them, then again...

After I had completed my task, I went onto the path and took inventory. Once I had made a complete list, I contacted the lieutenant, but before I got a chance to start, an RTO from the third platoon came on the horn screaming for a medivac. I shut up, released the button and listened.

One of their guys was exploring a cave and spotted a gook lying inside. As he approached, the gook turned and fired hitting him in the chest. Another guy was behind him and opened up on the gook, but before he killed the gook, he took a bullet in the foot. The man shot in the chest was in pretty bad shape. He died before the medivac came out. Another one lost to the hill. The dust off helicopter got the dead guy and the wounded man out without further incident.

In the meantime we packed up the items and headed back to our position. If anything was happening, we wanted to be back together. Before we left, Rags placed a white cloth over the spot where all the machine gun bullets had hit that rock. This would serve as a marker for tonight.

We had decided to stay in the same spot another night. Not the wisest idea, but a possible strategic move. The gooks would not expect us to be in the same spot.

We got back to the ravine very quickly. I laid the items out and checked my inventory. It was okay. I called the lieutenant back and started reading the list of items found.

"... he was killed by an M79 grenade. He was approximately 17 years old. He had one AK47 with banana magazine, extra rounds, one pound of tobacco, one pack of Vietnamese cigarettes, ten pounds of rice, potatoes, a cooking pot, a bag of white stuff..."

As soon as I said that I saw the guys waving at me and heard the CO come on the horn.

"Kid make damn sure that white stuff gets on the bird to be sent back in. You understand me?"

"Roger."

I looked at the guys bewildered. One of them explained.

"Opium dummy. Now we gotta send it in. Dumb fuck."

I continued with the list.

"...one gook web gear. He was carrying a wallet with some papers and about six hundred Piestras. He had a GI canteen cup, one new gook poncho, one new sleeping gear, a gook first aid kit, a watch, a pair of Ho Chi Minh sandals (the sandals had been strapped to his web gear, he was walking barefoot to move more quietly) and one silver spoon."

That completed the list. The CO came back on the horn.

"You guys did good. I was beginning to wonder. Make sure everything gets on the helicopter, especially the AK47 and that white stuff. Good work, out."

I remembered one more thing. I called the lieutenant back and told him that the gook's nose was plugged. It was stuffed with white powder.

The opium helped the enemy to endure the punishment we were dealing them. The guys explained that while high on the stuff, the gooks could, and would be driven to endure or do just about anything. One guy knew of an instance where the gook had taken several direct hits and still kept coming. The gook did not stop until his body actually stopped functioning.

It was not uncommon to find opium on gooks. The bag (of opium) looked and felt like about ten pounds. I wondered silently what that would be worth on the streets of Chicago. I was reasonably sure it was pure. Jack had told me once that I would probably find the "stuff" back in the world much weaker. It would probably take twice as much to get high, and I would pay dearly for it. In Vietnam, it was free and it only took a couple of joints to start feeling high.

Before the helicopter came out and picked up the items, we had taken some pictures of the AK47 for the folks back home. One thing about the AK47 was once you heard it fired, especially if it was at you, you never forgot the sound it made.

The AK47 had a particular crack to it. It appeared to be more of a snap, rather than a popping sound. The AK47 sounded the same as any other rifle when it was fired single shot, but when it was fired on automatic or multiple shot, it gave sort of a high pitched cracking. The more shots fired consecutively, the higher the pitch.

Someone had said it was due to the way the barrel had been fit onto the stock. Someone else had said it was because of the thirty round (banana) magazine they used. The thought was it was actually the magazine giving off the snapping sound. I suppose no one really knew for sure. Maybe it really wasn't different, but I swear it was a sound I have never forgot.

The one flaw of the AK47, if you could call it that, was the wood stock. It was not very compatible with the jungle climate. The wood, if not kept dry, would start to rot, or at least loosen, causing the rifle to become inaccurate. However, using the quick kill theory, it was not necessary to be accurate, just effective. The AK47 was very effective.

In comparison, the M16 appeared to be a much superior weapon, although many thought the AK47 was the better weapon. The M16 had a plastic stock and fired a smaller round. We used a twenty round magazine, but only filled the magazine with eighteen rounds, so as not to put too much pressure on the spring. We also emptied the magazine periodically to make sure it was still functioning. At times, we did this by firing up the rounds. Other times, by just ejecting the rounds out. Both the spring and the magazine were very susceptible to rust.

We did have a thirty round magazine, but very few used it. The thirty round magazine was slightly angled and was called a banana magazine. Each man usually carried ten to twelve twenty round magazines, giving him around two hundred rounds of ammunition. The trick was not to fire it all up at once, but to wait for your target.

The M16 was not the only weapon we used. Of course we had the M79 grenade launcher and the M60 machine gun. However, some chose other weapons. The sniper we had with us on the hill used an M14 for the better target shot. The M14 also had a wooden stock. Another guy in our unit used an M16 with a silencer mounted on it. It had the effect of a twenty-two for sound. I wasn't sure why he needed a silencer, but he was content with it.

# | 170 |   THE PROTECTED WILL NEVER KNOW

Quite a few guys had their own private pistols that they had brought from the states. I heard of guys carrying M1s and grease guns and other types of rifles they had brought or requested. Hell, whatever turns you on (or keeps you alive).

The army gave you very little hassle about bringing your own weapons to Vietnam. The theory was, whatever you were comfortable with.

The classic private gun story I had heard involved an Australian (or American), fighting on the side of the Viet Cong up near the DMZ (Demilitarized Zone). This person carried two revolvers in holsters strapped to his leg, just like the old west. He wore what appeared to be a cowboy hat and could be seen riding standing in a jeep. The hat gave rise to the suspicion that he was Australian.

There was a five hundred dollar bounty on his head. The bounty would be paid to either an American or an ARVN, but you had to produce the body with guns and hat. Another rumor circulated that some general was offering one thousand dollars for the guns and hat... screw the body.

The rear had sent a different machine gun out with the recovery helicopter. We were not able to fix the one we had. Rags had told the CO he was going to leave it for the VC. The VC couldn't use it on us anyhow. The CO got the hint and ordered another one for us.

Along with the gun was clean ammo. Besides the five hundred the M60 gunner carried, the other guys in the squad carried at least one hundred rounds each. This usually made a little over one thousand rounds available to the M60 gunner. None of the RTOs carried any rounds. We had enough extra weight. Most guys carried the belts of rounds across their chests similar to the way the Mexican bandits would in those westerns. That is how the ammo would get dirty. Even though the ammo we had looked okay we sent it in anyway.

Hopefully the M60 machine gun would be effective next time. Rags was very doubtful. He requested and received permission to test the new gun. It worked quite well. Rags was still doubtful.

Night was falling, we would know soon enough if the gun or our decision to stay in the same spot would work out.

# LET'S DO THIS AGAIN

**WE STARTED TO PREPARE FOR THE NIGHT.** We dug the foxhole a little deeper, mounted the machine gun with the aim fixed toward the spot on which Rags had left the white cloth and waited for dark.

Rags had spread the word that if we spotted the gooks again, he wanted a couple more guys to join us for additional firepower. We did not have any idea how long we would have to wait or even if they would come back.

We stayed with the other squad and set up much the same as we had the night before. It was a battle of wits now. We would not have to ask for permission this time. It was assumed we could handle it or at least Rags could.

The wait was not long. Shortly after dark, Rags was checking the spot with the starlite scope. He whispered to get ready. I motioned for the others to join us. We sat there listening to Rags directing.

"I want that gun aimed right at the cloth... don't shoot 'till I give the word... ya got the 79 ready? When I say so, keep firing 'till I tell you to stop... you two start firing after the machine gun points out the area… Kid, stay on that horn and don't get in the way… okay, there's one of them… hold it… there's another… wait, not yet… damn, the other one went back… wait here he comes… There's three now… wait, wait, he's walking into the area… he's going by the… he picked it up. The gook picked up the fuckin' cloth… fire, fire, fire that mother fucker up… I don't believe it, the gook picked it up…"

In an instant, the quiet of the night was met with a maddening burst of gunfire. The machine gun never sounded better. All hell was breaking loose. The spot was aglow with bursting and ricocheting rounds. The guys were covering every inch of the area with their shooting. The air was filled with gun smoke. It was a performance to be proud of.

I could barely hear the CO calling on the horn. He wanted to know what was going on. I tried to explain about the cloth and about the gook picking it up, but he wasn't sure he understood what I was talking about. The CO thought we had hit a platoon of gooks the way we were pouring it on. The CO said we should slow down, that we had already given our position away.

In the next instant everything stopped. I told Rags that the CO was on the horn. Rags said, "tell him thanks for the gun". I relayed the message to the CO. The CO responded. "Go to hell" and signed off. I relayed the message to Rags. He just smiled and went back to watching the action.

The rest of the night was quiet. We kept a watchful eye on the spot just in case the gooks were dumb enough to come back. They were not. In the morning we would have to go back and see if we were successful again.

After we had breakfast, the three of us went out again. Rags, the cherry boy and myself headed for the spot. Rags thought that since we had put on an exhibition last night, three would be plenty. The rest stayed back watching and directing our course. They could cover us if it was necessary.

The three of us proceeded down over and up toward the path we had found the day before. In no time at all, we were at the rock on which Rags had left the cloth.

The rock was covered in blood. We had hit something. There was fresh blood on the grass and something else, possibly flesh or muscle. The mess was caked in one spot then started trailing up off the path.

As we rounded the rock, I looked in where we had left the body. The dead gook was still there. The maggots had broken through the chest. I turned away, but something struck me as being strange. I looked back immediately. I saw what bothered me. The undershorts were missing. I knew I had left them on the day before. I called out to Rags.

"Hey man, someone took his shorts."

"What?"

"His shorts, I know I left them on, but they're gone now."

"What the fuck do you think that white cloth was that I left on the rock?"

"You mean...?"

"Yeah, what the fuck, he didn't need them. Hell, it was the only thing I could find. I wonder what happened to them. Hey call them, tell them we're going to follow the blood for awhile."

I reported in and we started up the path. The new guy was leading. We found the shorts. They were full of blood and whatever that other stuff was. The blood was thicker in spots and there were indications of someone being dragged. We definitely hit someone, but where was the body?

We reached a cliff and as the new guy started to peer down he was met with a burst of gunfire. The new guy jumped back and Rags and I hit the ground. I called over to our guys that we were in contact and could they see anything. They could not. We fired a few bursts back and started to inch our way out of there.

We were not in a position to mount an attack, so a retreat would be more appropriate. Just for cover we lobbed a few grenades down and made our break. As we passed the rock, I called over for the guys to open up on the area. We were moving fast and would be clear in a few seconds.

Once we got back to the ravine, the CO informed us he had ordered fugas to be dumped on the area. We sat back and waited, watching to make sure no gooks escaped.

The first Chinook helicopter was arriving and about to move into position when one of the guys noticed rifles protruding from some rocks. I called the helicopter and told him to pull out. He did instantly.

Rags told the M79 guy to start pumping the spot. Hopefully it would keep the gooks down long enough to drop the gas. The plan worked. Ten fifty-five gallon drums of fugas were dumped over the rocks. The drums exploded on impact. A huge cloud of black smoke erupted.

A second Chinook helicopter was flying into position. Another ten drums were dropped. The two helicopters left heading back for more.

We watched, but could not see through the black smoke. Rags yelled to put more grenades in there just in case. The idea was for the fugas to seep into the caves and hopefully drive the gooks out in the open or burn them up in the caves.

In just a short time, the two Chinook helicopters were back with another load. The first helicopter dropped his load straight down. The second helicopter got a little fancy. The helicopter came in right over us then dipped down and went at the hill in an upward sweep. The effect was perfect. The motion caused his load to spray across the area.

One of the drums bounced out of the pack and did not ignite with the rest. As the drum was making the second or third bounce, the door gunner on the first helicopter, that was still hovering nearby must have see the drum and opened up on the bouncing drum with his machine gun. The drum exploded in mid-air sending burning gas flying everywhere. The whole scene resembled one of those spectacular firework shows you see after a sporting event back in the world. It was quite impressive.

Like icing on the cake, a few rounds of heavy artillery were directed to the area. Needless to say, it made us quite nervous having those shells flying over and exploding so close to us. We decided to find a new spot. Besides, three nights in the same spot was looking for trouble.

We found an area that was about the size of a football field. The area was completely enclosed in shrubs. The grass inside looked as if someone had recently mowed it. There were two paths leading into the area and both were easy to guard. A high spot provided a great spot to post a lookout. An abundance of shade provided an opportunity to get out of the sun for awhile.

The best feature of all was in the back. A stream coming down from the hill that fed into this place. As the stream passed through to the rice paddies, the rocks formed a small section of water about three feet deep and five or six feet wide. We could take a bath.

This place was almost too good to be true. As the land dropped down to the valley it formed into a series of rice paddies. That side would be easy to guard, as it was all wide-open area.

We received word that our other squad would be joining us because the 173rd had finished their sweep and were moving back into position. All three of the ambush squads would be back together. The lieutenant and the sergeant were still up on the top of the hill. We could really defend this place now.

The squad joined us late in the afternoon, with just enough time to take a bath before we would have to set up for the night. We sent some guys out to look for a place to setup another ambush. They had found an island in the middle of the paddies. The spot was in line with the hill and the village. We headed there.

Rags and Lumpy, the guy in charge of the squad that joined us, were at first going to divide up the men and pull two ambushes, but then decided to stay together. It was too big of an island. Actually, there really wasn't another spot.

I had just fallen asleep when I heard the explosion. More explosions then screaming followed the noise. Two gooks had walked right into us. One gook was laying about ten feet away while the other gook had jumped into a clump of bushes that were now being spayed with gunfire. I jumped into the foxhole with Rags and grabbed my radio. I had drawn the forty-five out and was trying to get a fix on the situation.

Right in front of us was the first gook, his body going through severe muscle spasms, in the last stages of life. Suddenly from out of nowhere, Lumpy came diving over us toward the gook. In an instant, Lumpy grabbed the gook's head and with one quick swipe, had slashed the gook's throat. A moment later, Lumpy was back in the foxhole with Rags and me. I looked at Lumpy as if he had gone mad. Lumpy explained that when he awoke all he saw was this gook crawling toward us and his mind told him he had to stop him.

Lumpy and I both noticed Rags slumped over holding his gut. Lumpy grabbed Rags and straightened him up. Rags held out his hand. It was full of blood.

Rags explained that the gook was holding a grenade when Rags spotted him and Rags had quickly blown the claymore, but the grenade had exploded outside the foxhole. Rags said his hand was still holding the detonator and had caught some of the blast.

I climbed out of the hole and fetched a bandage. Lumpy wrapped the bandage around Rags' hand, but the bandage wasn't too effective. Rags was still bleeding heavily. We did not have the medic with us. The decision had been made that since we were so close it would not be necessary, besides the medic couldn't be with each squad at the same time. Ironically we had all joined together.

Lumpy decided he wasn't going to take any chances. He instructed me to get a hold of the CO, to call in a medivac and if the CO was sleeping to do it myself. I called and the CO had been listening to the action. He was already on top of it.

We hurriedly gathered all the loose gear together so that when the dust off came in, the helicopter would not blow everything away. The dust off helicopter was there in a few minutes. We quickly got Rags on.

While we were doing so, the wash of the blades set off some of the trip flares. Instantly we dove for the ground, not sure if the bird had done it or if there were more gooks out there. When nothing happened in the next few minutes, we regrouped and prepared for the rest of the night.

Lumpy decided to pump M79 shells out all night at different intervals and different spots. He also doubled the guard shifts for the rest of the night. He doubted anymore would come, but he didn't want to take any chances.

The next morning we searched for the other body. It was still in the bushes. Someone dragged it out. I was not stripping either one this time. We didn't have to. Intelligence had enough, just the gear and weapons this time. We left the two bodies and headed back to the place. Once inside we relaxed and cooked breakfast.

I intercepted a call from a helicopter pilot looking for the sight of the kill. I answered that we were right by it and could I be of assistance. The pilot asked if we would mark the spot. He had someone on board who wanted to see the spot. A couple of guys and I went back to the island, popped a smoke grenade to mark the location. The pilot acknowledged the smoke, bringing the helicopter almost down on it.

As the helicopter came closer the pilot angled slightly and we saw a first lieutenant hanging off the side taking pictures of the bodies. I looked at the other guys, but they just shrugged their shoulders. They didn't know either. After the lieutenant had enough, the helicopter took off and the pilot said thanks. We headed back to the place. No one there knew what that was about either.

The cherry boy had also taken some shrapnel from that grenade, but not enough to do any real damage. Still, Lumpy thought he should have a tetanus shot. Again he instructed me to call the CO to have someone come out and give the new guy a shot. The cherry boy thought it was silly. He was arguing with Lumpy.

"…com'on man, I don't need one."

"Listen cherry, it won't hurt for Christ sakes."

"Why do you keep calling me cherry. I got shot at yesterday and hit with shrapnel last night. What does it take?"

"That's right, sorry man. I forgot. What's the name?"

"Steve."

"Okay Steve, take your shot like a good little boy now."

Steve threw his helmet at Lumpy in jest, but Lumpy was already scampering away, laughing up a storm.

The helicopter arrived with the medic to give the shot. We brought the bird right down inside our place. It was a perfect fit; this place was really all right. Two people got off the bird and the bird took off. The pilot said he would be back in about a half hour to pick them back up and disappeared in the sky.

The medic and the other guy gave Steve the shot and were giving us the third degree about the ambush and particularly the kill. Neither one of them had ever been in the boonies before. Lumpy was trying to be cordial while at the same time annoyed with these guys.

"What did you do with the bodies?"

"We left them."

"You left them? You mean they're still there?"

"Yeah, why?"

"Can we go see them?"

"For what?"

Lumpy was starting to really get annoyed now, but the two guys kept pressing.

"We've never seen dead gooks before."

"What are you guys…? Morbid or something?"

"No man, just want to see them. Do they still have their ears?"

"What?"

"The ears, are they still intact?"

"Of course… yes… hell I don't know. What the fuck are you talking about?"

"Don't you guys collect ears for souvenirs? All you have to do is wrap the ear in saran wrap and it keeps."

"You two fuckin' crazy or something? Who the fuck wants an ear?"

## THE PROTECTED WILL NEVER KNOW

Lumpy was beyond annoyed now. You might say he was bordering on decking these two guys. Others were starting to gather around us. The two guys noticed the others starting to gather and quietly spoke.

"We do. Are they still there?"

"Yeah, I guess so… Help yourself. Fuckin' idiots we got here."

"Will you guard for us?"

"Hell no, you want the fuckin' ears you get them yourself."

"Alright, where are they?"

"Right out there, and if you get shot or killed out there we're not coming to get your sorry asses."

Lumpy pointed to the island. Lumpy also posted a guy to both watch and make sure they got back. The two guys headed for the bodies. In a few minutes they were back, each brandishing an ear. The helicopter was approaching to pick them up. We were glad to be rid of both of them.

The lieutenant called, he was on his way to join us. The CO wanted the platoon back together. We would be pulling more of a guard position rather than an ambush for awhile. We had been down here three nights and all three had been productive. We should take it easy for now.

Shortly after the lieutenant arrived, we noticed a helicopter flying overhead. The bird kept circling and circling. The CO called and explained that it was a visiting general looking things over. The general had spotted us lying around and suggested that since we had a break in the action we should take the opportunity to zero our rifles. Zeroing one's rifle meant making sure the sights were lined up and you were able to hit your target. In other words target shooting. When I put the question to the CO in not so many words he reminded me the general had suggested it. In simple terms it was an order. What the hell, why not?

A few seconds later, the CO called back on the horn. He informed me the higher ups were sending a helicopter out with body bags to pick up the two dead gooks. That really threw me. I asked the CO why, for what. All the CO said was "the general suggested..." I knew... I knew we had a problem.

I looked for the lieutenant. First I told him about having to zero our weapons, to which he thought I was crazy. I waited for his laughter to subside, before telling him about the second problem.

"Sir, they want the bodies."

"What bodies?"

"The gook bodies from last night."

"So what, let them have them. Who wants them anyway?"

"The CO said higher ups, must have something to do with that general flying around."

"So, give them the bodies. What's the problem?"

"Sir, I don't think you understand..."

"What's to understand? They want the fucking bodies, let them have the bodies."

"But sir..."

"Jesus kid, what's the matter with you?"

"The ears."

"The what?"

"The ears. Those medics that were here this morning cut the ears off."

"They what? Why? What the fuck's going on here?"

"Sir, they wanted them for souvenirs and we... well, we let them have them. Now what do we do?"

"Christ, we can't let them have those bodies without the ears. Let's see if the people in the ville' want to bury them. We'll tell the..."

"We tried that sir, they didn't."

"Why not? Didn't they want... well, what did they say?"

"Who knows? No one here speaks Vietnamese. We couldn't understand what they were saying and they didn't understand what we were saying and... Well, hell, no one got the idea so we left the bodies out by the island and when those guys came this morning who cared if..."

"Ah shit, why me? Now what are we gonna do kid? All the fucking bodies in Vietnam and they pick the two without ears."

We had no choice, we had to risk it. If anything came of it, we were definitely going to put the finger on those two medics. The helicopter flew over and dropped the body bags out near the bodies. A few guys went out, loaded the bodies into the bags and waited for the helicopter to pick them up. We all watched the helicopter take off and head back toward base camp. Later on in the afternoon we saw the helicopter circling around. I recognized the pilot's voice from the earlier pickup. The pilot was looking for a spot and checking with the forces in the area.

I told the lieutenant I didn't think it was the general's bird. However, we had been firing our rifles all afternoon using just about anything for targets, just in case.

I had concentrated on the forty-five, shooting it with both hands and practicing drawing. I had oiled the holster some more and was developing a rather nice draw-and-fire move. I tried different ways of making it accessible to my left hand, but without success. It would all come down to how I was holding the horn at the time I need the pistol. Hopefully, I wouldn't be in that position.

We were unconcerned about the helicopter. If he could see us, and we were sure he could, he would see we were doing what he suggested. The lieutenant and I continued to watch the bird flying around, as did others. The helicopter was proceeding toward the top of hill 474 and further over to the far side. We noticed something big and black drop from the bird. Then another big black object dropped from the bird. After that, the bird took off flying high and back toward base camp.

The CO called and the new guy monitoring the horn yelled to the lieutenant. "The CO says the bodies have been taken care of, everything's fine." The lieutenant and I looked at each other. He breathed a little easier. We went back to our shooting.

The lieutenant had ordered twice the usual ammo to replenish the ammo we were firing up for target practice. With the sky clear, the lieutenant called a halt to the firing, instructed everyone to clean their weapons and call it a day.

The rest of the night was peaceful. We would rest easy tonight. In fact, we spent the night right in our place. It was as good a spot as any. Just as easy to defend at night as it was during the day. The lieutenant decided to make this our permanent spot for the next few days. Whatever we had to do we would do it from here.

Fully relaxed for the first time in four days, I slept well, very well.

# ENJOYING LIFE

THE NEXT DAY WE WERE DUE FOR RE-SUPPLY. The lieutenant requested the helicopter be brought into our resort, as we had renamed the place. This time I had directed the bird in.

We dug into our supplies. Somewhere along the line I had been designated the mailman and the distributor of the sundry box, probably because I was now the lieutenant's RTO.

We were re-supplied every three days, which meant we received mail every three days, but again, because of our separation from the main group, sometimes mail would not come for six or nine days. Mail was always welcomed, no matter how late it was. As was the custom, I passed out the mail first. While everyone was reading the mail, I would go about the distribution of the sundry box.

The sundry box contained ten cartons of cigarettes, shaving cream, razors and blades, shoelaces, toothbrushes and toothpaste, candy, cigars, pipe tobacco and other odds and ends we could use.

The main thing we had to distribute was the cigarettes. There were six cartons of regular cigarettes, two cartons of menthol and two cartons of non-filters. Although there were usually twenty-five of us in the platoon and only one hundred packs of cigarettes to go around, we still had plenty.

I smoked regular, of which I usually was able to keep a carton for myself. I was also able to keep four packs of menthol to break the monotony. That gave me fourteen packs of cigarettes to last three days. I was usually down to two or three packs by re-supply. While we were moving my cigarette consumption was greatly reduced, but when we were lying around, I was never without a cigarette in my hand. Often times I would light the next one off the one I had.

The lieutenant was a pipe smoker and usually had his own tobacco sent to him from the States, but occasionally he had to use the tobacco in the sundry box.

With the mail and sundries distributed, we next attacked the c-rats. There was more than enough to go around. In fact, some of the less desirable foods were just thrown away. We would dig a pit, throw our garbage in it, then burn it.

Every once in awhile, someone would throw an unopened can into the fire. As soon as the can had gotten hot enough, it would explode, sending food and smoldering garbage through the air.

Cases of C4 explosive had been sent out for us to plant in the hill, more so to give us something to do rather than for any real need to blow up the hill. Every morning we took turns trekking up the side of the hill, planting sets of charges, then trekking back and detonating the latest set. The whole ordeal became a game of who could plant and set off the most spectacular charge.

The third or forth time I went out, I was with a guy that was going to set off a blast to end all blasts. We had ventured quite a ways up and had reached a small ledge. The ledge had indications of a bonfire and sleeping arrangements. Upon further examination we discovered that the stream we were enjoying passed through this area, apparently originating somewhere up in the hill.

The stream was only a foot wide and about a foot and a half deep, but the water was ice cold (cold in terms of the jungle heat and humidity). Around the stream were a few shrubs on each side creating a partial shade to help keep the water cool. We all drank from the stream, enjoying its freshness. We had been filling our canteens and drinking the stream water instead of the water sent out from the rear, but our water was not as cool.

While we sat, both taking a rest and negotiating the layout of the charges, one of the guys called for us to come and see what he had found. When we arrived at the spot where he was standing, we all saw what he was swearing about.

Right in the stream, with the water washing over it was a gook body. From the condition of the skin, it appeared like the body had been there a long time. We dragged the body out and pushed it off a ledge, down the side of the hill. We shrugged our shoulders and chalked it up to experience. Someone joked about the body giving the water the extra flavor we had tasted. With moans of disgust, we came close to throwing him off the ledge too.

Afterwards we went back to setting the charges. The guy who was running the show kept counting off steps, checking his bearing and directing where and how many pieces of C4 should be placed. We had all carried quite a bit up with us and he was being very particular about how we should set them. We kept asking if we weren't placing quite a bit around, but he just kept shaking his head no and went right on directing.

We had been up on the hill for over three hours while

everyone else had finished in an hour or less. The rest of us kept nagging him to go back, saying that was enough charges, but he wasn't listening. He kept repeating, "...Just wait and see... just wait and see". Finally we finished and headed back to the resort.

While the rest of us dropped down to relax, the guy went about attaching all the detonators he would need to blow his charges. When he yelled he was ready, I grabbed my horn, which was still strapped to my back. I had been too tired to take it off, and called over the air. "Fire in the hole... fire in the hole." That was the signal that a controlled explosion was about to happen.

I yelled and motioned for the guy to go ahead and blow it. He yelled back for all of us to watch, that this would be great. With all of us looking on, he hit the detonators. Nothing happened. He hit the detonators again, still nothing. Everyone relaxed and turned back to what they were doing. The explosion would not happen. I was about to spread the word that it was a dud.

In the next instant, the whole side of the hill erupted in one gigantic explosion. We were being showered with dirt, rocks and anything else that had been caught up in its force. The whole resort was becoming engulfed in one huge dust cloud. Before the dust had a chance to settle, the CO was on the horn.

"What the fuck is going on down there?"

I was still coughing up dust and could barely answer that I would get back to him. The dust had settled on everyone's face and arms and hands. We all resembled miners that had just come out of a mine. The lieutenant was yelling at the guy.

"Are you crazy or something? What are ya trying to do, kill us?"

"Sorry sir, it didn't work."

"What the fuck do ya mean, it didn't work. Ya nearly blew the fuckin' hill away."

"Not that, the sign."

"What sign?"

"The peace sign. The explosion was supposed to form the peace sign on the side of the hill for us to see everyday."

"Are you fuckin' nuts? You blew up half the fuckin' hill. All we have to do is look at ourselves. It's all over us."

"But it was going to be the blast to end all blasts."

"It was, you fuckin' madman."

"I'll have to do it again."

"You what? Ain't no fuckin' way baby. Those fuckin' charges are going back in. That was the blast to end all blasts, because there ain't gonna be no more. If I let you guys keep playing with that shit, you'll blow us away instead of the fuckin hill. What ya think…? This is the forth of July for Christ sakes? That's it, get ridda that shit, burn it if you have to, but get ridda it. What the CO say, kid?"

"He wants to know what's going on down here. I think you better talk to him."

I handed the lieutenant the horn and listened while he tried to explain to the CO what had happened.

Everyone took turns going to the stream to take a bath, to wash all the dust off. The hill was still settling from the explosion and there were still small clouds of dust in the air. We watched as dirt washed into our stream and back out again. Fortunately, we had a continuous flowing stream.

Now that the charges were gone, we had practically nothing to do. We had been playing cards off and on, but now we were playing all of the time. Once in awhile we would have to scout the area, just for something to do. We had not seen or heard anything since that last ambush.

Toward the end of the afternoon one day, we got a call that

we were being flown to the other side of the hill. A LRRP team had been working there and had made contact. During our lull, I had practically taken over as the lieutenant's RTO. The new guy had been pushed on the sarge and the professor had become one of the two extra RTOs. However the professor was the senior and still had the privileges.

We went out to the island to meet the helicopters. There was enough room on the island to bring two helicopters down at once. The lieutenant always rode the first bird and I climbed in with him.

We landed on a grassy knoll on the backside of Hill 474. We were off in an instant, waiting for the others to land. When they didn't, I called to find out what had happened.

We were informed that we were set down in the wrong spot. The rest of our platoon had landed further over and since it was too far to walk, the helicopter was coming back to pick us up. I informed the lieutenant of this and he just shook his head in disbelief.

The rest of our platoon was sitting and relaxing when we finally arrived. They rode us about taking so long. The lieutenant was already quite mad about the whole situation.

Upon speaking with the LRRP team, we discovered that they had not made contact. They had only suspected enemy infiltration in the area and had asked for help. The lieutenant was now even madder. He knew the higher ups were just trying to keep us busy.

It was getting close to dark so we started to look for a place to set up, but we could not find one. We were halfway up the side of a small ravine. Our platoon was strung out, instead of in its usual circular setup. The steepness of the ravine forced us to sleep almost straight up. Guys were tying themselves to trees to keep from sliding down. It was one of the worst setups we ever had.

In the morning, we searched the area and could not find a thing. There was absolutely no trace of any activity in the area. When the lieutenant pressed the leader of the LRRP team for more details, he admitted that they were flown in just before us and that was why the lieutenant's helicopter had been diverted.

The lieutenant, needless to say, was not in the best of moods. He had me get the CO on the horn. When I did, the lieutenant snatched the handset from me and informed the CO of our findings, actually that we had found nothing. The CO instructed us to head back to the resort and if we left now we would make it by late afternoon. The lieutenant threw the handset back at me. He was understandably upset.

The lieutenant called the sarge and the squad leaders over to explain the situation and work out the best way back. I overheard the lieutenant and the sarge in a heated argument over which way they should go.

I suppose with the hassles he had been through and his dislike of the sergeant anyway, the lieutenant had had enough. The lieutenant put the sergeant at attention, something you never ever do in the boonies, mainly because it identifies who the officers and leaders are, which to anybody watching would be the prime target. Officers were number one on the list to get blown away by the enemy.

The lieutenant was now spitting out orders to the sergeant about how he was in command, that he, the lieutenant, would make the decisions and how he was tired of being belittled by the sergeant every time he made a decision. The lieutenant kept up a steady barrage of orders, saying that from now on

## THE PROTECTED WILL NEVER KNOW

things would be different and asked the sergeant if he had made himself clear.

The sergeant, true to form, had taken the verbal abuse without flinching. When the lieutenant released him, the sergeant did not even retaliate. He just went back to what he was doing. All the sergeant had said during the whole ordeal was a few yes sirs.

The lieutenant headed straight toward me. As he came closer, he started reaching out and I started to duck, thinking he was going to hit me. The lieutenant grabbed the handset from me with such force that the handset and my hand holding it, went up to his ear. As he spoke into it, I slowly slid my fingers from around the handset. The lieutenant noticed and pulled the handset away from his ear long enough for me to remove my hand. I thought I detected a short smile as he returned the handset to his ear. I backed away. I had never seen the lieutenant this mad and I wanted no part of it.

Several of the other guys had been in earshot of the ordeal and had stood quietly throughout and remained so afterwards. The squad leaders had quietly backed away as well.

We worked our way around the side. As we reached the front side, we hugged the base of the hill, so as not to walk through the open rice paddies.

The usual way we humped (walked), would be a point (lead) man first, a backup behind him, the lieutenant, myself and the machine gun behind us. We had three M60 machine guns, three M79 grenade launchers and four radios. All of these would be interspersed throughout, all the way to the back of the line. At the rear, working backwards would be a rear guard, his backup, then the sergeant and his RTO.

The lieutenant had the front and the sergeant the rear. That would keep them both separated for the day. Hopefully by the time we got back they would both be cooled off. I was sure the sarge was not happy with being told off in front of the men.

As we approached the island, the lieutenant stopped and pulled me off to the side with him. He instructed some men to continue on, but to first make sure no one had moved into our resort while we were away. The men dropped their gear by us so they could move more freely.

We watched as the other men filed past us. The lieutenant stopped the men when about half had filed past and instructed them to spread out more. As they did so, the lieutenant was preoccupied with his map and I had dropped my gear, holding onto the horn, monitoring traffic and watching the men spread out.

Neither of us was aware of the sergeant's presence until he spoke.

"Tonight mother fucker, tonight you pay for that."

I looked up at the sergeant. He was standing there in full gear facing us. He was pointing his rifle right at the lieutenant and simultaneously twisting the safety switch back and forth, engaging and disengaging the trigger mechanism. I stared in horror, mostly due to the fact that I was right behind the lieutenant and if the sergeant slipped and fired, it would most assuredly go through him and right into me.

I slowly sidestepped, waiting for something to happen. The sergeant left as quickly as he had come. The lieutenant staggered toward me and feebly reached for the horn. My hands were saturated in sweat. The lieutenant noticed. He asked if I had seen what just happened. I said I had. The lieutenant called the CO. I thought back to what Jack had said about hassling someone with a loaded rifle. I understood what he meant.

Will the real enemy please stand up?

## | 196 | THE PROTECTED WILL NEVER KNOW

When the lieutenant finished explaining to the CO, he called the sergeant over and handed him the horn. The sergeant, acting as though nothing had happened, took the horn.

I watched the expression change on his face, while he listened. I could not hear what the CO was saying, but I was sure it pertained to what had just happened. The sergeant handed the horn back to the lieutenant. The lieutenant handed the horn back to me a few seconds later and mumbled something about a helicopter coming out.

The lieutenant decided to bring the helicopter down on the island rather than the resort. The bird landed, the sergeant got on and the bird took off again. We all headed for the resort. The word got around about what had happened. It was interesting to note that most were glad to see him go.

The next morning a new sergeant, as well as some new recruits, joined us. The new sergeant was a twenty-year man. The story was that he had been up to E7 or sergeant first class, but had been busted down to private for associating with a whorehouse in Japan. Since then, he had made it back to E7. The new sergeant and the lieutenant hit it off right away.

We went back to doing nothing again, except playing cards. Most of us did not draw money, we had it sent home instead. We would draw a few dollars just for pocket money, but not enough to stake a good poker game. We were floating IOUs all over the place. The IOUs were as good as money. Periodically we would check to find out who owed who and consolidate the IOUs. It was not uncommon to meet a five-dollar bet with a ten-dollar IOU and remove a five-dollar bill from the pot. Whoever won the pot, won the IOU.

After another week went by, the CO decided to join us in the resort. He left the second platoon up at the top of the hill. He said he needed a bath and a rest. The second platoon had been sitting on the top doing nothing also. In the next few days they would join us as well.

Eventually we would have the whole company down here. The resort was big enough. By the end of the week they were all down. We would take turns going in for twenty-four stand-downs to a beach area the 173rd had set up back at LZ English. There would be plenty of beer, hot dogs and hamburgers.

We took our turn. It was another white-sand beach. They did have everything there. That night a bunch of us gathered in a large circle, even some guys from the 173rd. The guy next to me had a box full of rolled joints. He also had several packs of menthol cigarettes. He would light a joint, take a hit, hand it to me, light a menthol, take a drag, hand it to me and so on. We never saw them again. They didn't make it around.

Someone had started a pipe going around to fill the gap. It was a custom made pipe. Actually, it was an M79 grenade shell casing that had a piece of bamboo stuck in it for a stem. The pipe had been appropriately renamed super bowl.

I don't know how long I stayed there or when I left, but in the morning, I woke naked on the beach staring at the sun. The waves were crashing against the shore. As I lay there taking it all in, enjoying the warmth of the sun, it took me awhile to remember where I was. I looked around, there were quite a few other naked guys sprawled out on the beach with me. Apparently we all went for a midnight swim.

This was probably the biggest contradiction of Vietnam. With the fighting going on throughout most of South Vietnam, here were these absolutely great white-sand beaches where you could forget where you were for a moment and believe you were in a resort somewhere. Actually Vietnam was an exotic locale where the rich went to play and big game hunting was a favorite past time.

Vietnam was a French colony before the last real war, dubya, dubya deuce (WWII) as we referred to it. It was not unusual to come across an old deserted French mansion or plantation in the jungle during our travels. You could see the beauty of the land in the countryside (jungle, boonies as we called it) as well as some of the old architecture in the cities.

While I sat on the beach in the warm sand watching the waves splash against the shore, the realization of where I was at disappeared for a moment and I let myself enjoy the serenity.

Someone yelled that it was time to go. We were told there would be a nice fresh cooked breakfast waiting. I looked over at the other guys starting to stir and surrendered the moment.

In fact the breakfast was not cooked until you ordered, so there was no need to hurry. They had plenty of coffee on hand, soft drinks too. The beer had been taken away. We were going back to the boonies that day, so no more drinking.

During the day we presented the Battalion Commander with the rifle we had captured on our first ambush. The AK47 had been cleaned up and had a plaque affixed to the stock. The rifle was his present from us, from Hill 474.

We went back to the boonies. We spent another week in our resort before we got orders to abandon Hill 474 for new regions.

We had been at Hill 474 about two months. Four down, eight to go I thought.

We would be leaving in the morning, to places unknown.

# ON THE MOVE AGAIN

**We were picked up on the island,** two helicopters landing at the same time. We climbed aboard. I was still riding with the lieutenant. The professor decided to work with the new sergeant. We headed toward the horizon.

We landed in a valley between two small hills. We were able to bring four helicopters down at once. The fifth came in after us. We would be working in platoon strength again.

After we were on the ground and the helicopters had left the area, we regrouped and headed toward the hill on the right. This hill was mostly vegetation and had very few areas of rock. In fact, it was so thick with growth that we had to cut a path through it. Even with that against us we were able to reach the top in just over an hour.

Upon reaching the top, we discovered that we would probably not find anything in the area. Looking down each side, we could see there had not been any traffic through here. We sent a couple of patrols out anyway, but their findings were negative.

## THE PROTECTED WILL NEVER KNOW

We climbed back down and headed up the other hill. Still nothing to be found. We broke for chow and searched for an area to set up in for the night.

We spent a couple of more days checking the area, but were unable to find anything. The next morning we were picked up and carried to another place.

We worked that area for awhile then moved on by foot. We wound up in a graveyard and decided that it was as good a place as any to break for chow. We moved onto an area of flatland near a stream and took our re-supply there. Fortunately for us we did. Our re-supply cans of fresh water had a very strange taste to them. One of the guys surmised that the rear had gotten the cans mixed up again and sent our water out in gas cans. They had done that before.

The lieutenant had asked for his forty-five back and I had hesitantly given it back. However, sometime later, the lieutenant had misplaced the forty-five and now it was gone. The lieutenant made a futile attempt at accusing me of losing it. He finally relented and contacted the rear. The higher ups sent the lieutenant a bill for five hundred dollars for having lost it. The lieutenant should have listed the forty-five as a combat loss. He was quite upset.

The CO, the lieutenant, some others and myself were summoned back to the rear to appear at the sergeant's court-martial.

Apparently, incidents of officer genocide had sprung up in the north, where our unit, the 101st Airborne Division, was based and the brass had decided to set an example with this case. In other words, the brass was going to "burn" the sergeant. We, the lower ranking personnel, did not think that was exactly fair.

At the court-martial, I testified that the incident was more of a reaction, rather than an actual attempt. Unfortunately, I had been the only witness, besides the lieutenant, and my testimony was crucial. A few of the other guys testified to the sergeant's character and his usual outstanding performance. I believe even the lieutenant did not want him crucified.

Whatever our feelings as well as the courts, the judges chose to be lenient. The sergeant was busted down to private, assigned to another company and subsequently given the M60 machine gun to hump.

My day in court had taken less than a few hours. I was back in the boonies the next day, but it wasn't soon enough. I had to shave, get a haircut, change clothes and shine my boots. After three months in the boonies, there was no hope for my boots. I borrowed one of the clerk's boots for court and for the hell of it I ordered a new pair.

The platoon was still in the area by the stream, having decided to work out of there until the lieutenant came back. The lieutenant rejoined us that afternoon. He sent out a squad to pull an ambush on a path they had found in our absence.

The squad had an element of VC walk into their ambush. We could see the light of the trip flare and heard the sound of the claymore exploding in the distance.

The next morning the squad came back in carrying some gear and an M16. The lieutenant questioned them as to what they had hit and who they had killed. The gear was a mixture of VC and GI, plus so called "black pajamas" he was wearing more than enough to identify him. ARVNs were just about carbon copies of GIs.

I picked up the captured M16 to look at it. The side was perforated with bee-bee holes. The claymore had made a direct hit. As I spun the M16 around in my hands, I noticed the safety had been turned to full automatic and was still in that position, with the magazine still engaged. One accidental mishandling of it and things could have been disastrous.

One of the guys from the ambush squad came over and took the M16 from me. He said he knew it was that way. The reason they weren't worried was because they had tried to clear it. When that had failed, they tried to fire it. Neither would work. He pointed the M16 in a safe direction and pulled the trigger. Nothing happened. Apparently the mechanism became nonfunctional from the destruction of the claymore. Whatever the case, the M16 was now harmless. He tossed it to the ground and walked away. I stood somewhat embarrassed at making an issue over nothing. I used the tip of my boot to push the muzzle away just in case.

We moved out in the morning. We were off to another area. We had received a call to help another LRRP team that had taken fire from a hill in an area close to us.

The helicopter flew into a covered area that was marked with smoke. I was looking at the ground, which was about five feet away. I was waiting for the helicopter to get closer to the ground. The door gunner next to me kicked me in the side motioning for me to jump. I looked at him like he was crazy. He pointed to the trees surrounding us indicating the helicopter could not go any lower.

I released the straps and jumped, pulling my ruck behind me. I hit hard and immediately sunk to my chest. My rifle went completely under water. The ruck splashed behind me sinking deep.

As I tried to maneuver, the lieutenant came crashing behind me causing me to go completely under. When I resurfaced I felt a stick pricking me on the back. I turned around and saw one of the guys from the LRRP team offering me the stick to help me out. I threw him my rifle, grabbed the strap of my ruck and pulled myself toward him. We both worked on getting the lieutenant out.

The helicopter had pulled up, hovering above us. The other three had not jumped. The guy from the LRRP team was talking on the horn. He motioned for us to follow him. The lieutenant and I looked at each other.

We had landed in a quagmire: grass, mud and water. We followed the LRRP guy out to an open area about twenty yards from where we had dropped.

The helicopter landed and let the other three off there. The three, upon seeing the lieutenant and I, obviously wanted to say something but the look on the lieutenant's face convinced them otherwise.

The other helicopters came in and landed, bringing the rest of the platoon with them.

We proceeded up a path on the hill toward the area in which the LRRP team had made contact. We approached an encampment, which had just been riddled with gunship fire. There was a gook body lying there. It was a woman dressed in black pajamas. There were also various supplies lying around, including bandages and other hospital type items.

We passed it all over for the moment. We wanted to pursue the path and trail of blood leading away from the camp. We did not get very far before we had to turn back. The going was too rough for us with all of our equipment. So we went back to the camp. A new guy stripped the body and collected all of the equipment. We headed back to the open area.

We worked this area for a couple of days before pulling out.

## THE PROTECTED WILL NEVER KNOW

We found a deserted village with four small hooches (grass houses) and we decided to blow them.

One thing of interest was found inside one of the hooches. We had found some ammunition. Upon closer examination we discovered a bunch of silver bullets. At first it was assumed that they were silver plated, but when the bullets were scraped down they proved to be pure silver. We all started looking for a white horse. Quickly the lieutenant felt that we had run that joke into the ground and he gave the order to detonate.

We were in and out of areas too numerous to mention, but we did a lot of traveling. We hardly found anything and the few things we did find were very old. Everywhere we went we left the numbers "101" burned into trees to indicate our presence. We were starting to feel like gypsies again. We were spending more time in the helicopters than on the ground.

We were next slated to be sent to the top of a hill that was supported by an ARVN unit. We would guard the top while they searched the hill. It wasn't too encouraging working with ARVNs, but at least it would be sort of a rest.

We landed without incident. Once on the ground we moved into position. The ARVNs left shortly after we arrived. We spent the day winding down, playing some cards and taking a re-supply.

The re-supply contained my new boots. A couple of the guys saw me putting them on. They came over asking if they could get new boots too. I answered that they could, but they had an ulterior motive. They each grabbed a boot and proceeded to grind it into the dirt, bending and jumping on the boots as they did. When the two guys had finished with their fun, they threw the boots back at me. I put the boots on anyway. They were still newer than the ones I had.

During this time several of us from Hill 474 received our medals. They were pinned on by the CO. We were given the cases and orders supporting the medal. I looked mine over.

**Awarded:** The Army Commendation Medal with "V" Device

**Date of Service:** 25 January 1970

**Theater:** Republic of Vietnam

**Authority:** By direction of the Secretary of the Army under the provisions of AR 672-5-1

**Reason:** For heroism in the Republic of Vietnam on 25 January 1970. Private Meyer distinguished himself while serving as a radio telephone operator in Company D, $3^{rd}$ Battalion (Airmobile), $506^{th}$ Infantry, during a reconnaissance in force mission near Landing Zone English, Republic of Vietnam. When Private Meyer's element was attacked with hand grenades and small arms fire, two of the platoon's three radios malfunctioned. Private Meyer handled all communication on the one remaining radio. He directed gunship and artillery support and maintained continuous contact with friendly elements in the vicinity. Discovering that smoke grenades were needed to pinpoint a pickup zone for medical evacuation helicopters, Private Meyer crawled through intense enemy fire to obtain them. His courage and professionalism greatly contributed to the success of the mission. Private Meyer's personal bravery and devotion to duty were in keeping with the highest traditions of the military service and reflect great credit upon himself, his unit and the United States Army.

I put the ribbon back in the case, folded up the orders. Later, I made arrangements to have the package mailed to my mom back in Chicago.

After a couple of days of lying around, the lieutenant got the itch for a good poker game. We gathered four other guys and headed for a clump of bushes to play. The whole company had been brought here and we did not want the CO catching us playing. We were not sure what his reaction to playing cards in the boonies would be.

We had been playing for about a half hour, when suddenly a familiar voice broke our concentration. It was the CO.

"What the hell is going on here?"

The lieutenant tried to speak.

"Ah... sir..."

"Ah sir shit. What the fuck do you think you're doing lieutenant? Do you know where you're at? This is the boonies lieutenant, not a fuckin gambling hall. You have the audacity to hold a poker game here? I'm really surprised at you lieutenant. I thought I had you figured better. I just can't believe that you would not only sanction a game, but play in it yourself. What surprises me most of all is that you did it right behind my back."

"Sir, I can explain... you see..."

"Forget it lieutenant. It's too late now. The damage is done. What's your story kid? What have you got to say?"

"Ah... do you want to join us...?"

"Damn, thought you'd never ask. What's the game?"

The CO crawled into a spot across from me. I could see the lieutenant still shaking. The initial shock had not passed. After a couple of hands, the lieutenant was back in stride.

It was my deal. I was dealing five-card stud, four up, one down. I had three cards out to everyone. I had a pair of queens showing, normally a good enough hand to win in five-card stud. The CO had two cards into a flush and another guy had two cards into a straight. The chance of either one catching was very slim, especially with only five cards. All five had to match to make it work. I was not worried, I bet five dollars, everyone called. My queens had not scared anyone.

I dealt the third card up. I gave the CO his third card into a flush and the other guy his third card into a straight. I gave myself the third queen. I bet ten dollars, the CO made it fifteen the other guy made it twenty. I made it twenty-five. Everyone but the three of us dropped out. The three queens were almost a sure winner.

I dealt the fourth and last card up. The CO got his fourth card into a flush the other guy got his fourth card into a straight. I bet twenty-five, the CO made it fifty, the other guy made it a hundred. I looked hard at both of them. They had to be bluffing. The odds of making either one of those hands were extremely slim. I borrowed two hundred dollars from the lieutenant and threw it into the pot. The CO, realizing I was out of money, called, so did the other guy.

I turned up my hold card. It did not match anything. All I had was the three queens. The other guy whipped his card over. He had the straight. My mouth dropped. As he reached for the pot, the CO caught his hand and turned over his hole card. The CO had the flush and quickly scooped up the pot and just about everyone else's money and IOUs. The CO now owned the highest sum of IOUs in the platoon.

As the CO walked away he mumbled something about playing cards in the boonies. We had learned our lesson. We would not do that again. The next time we got an urge to play, we would definitely not ask the CO to join us.

We played cards after that, but confined it mostly to tonk. The stakes weren't as high.

After a few more days it was decided we should pull out and start checking the area also. We had just barely left the top and started down when we took fire.

The initial burst exploded in a clump of rocks right in front of us. Our point man yelled back that he had been hit. He was just a few feet away from me. I looked over toward him. He was trying to get his ruck off, but it had become tangled on his arm. I could see why. He had a gaping red hole in the back of his upper arm. I could see the pain in his face as he struggled. The guy behind him grabbed the back of his ruck, pulling the man to the ground. Although it added to the pain, it was better than him standing there in the open.

By this time the medic had reached him, applying first aid to the wound. A dust off had been called. A squad of the second platoon was making its way past us toward the gunfire. The squad had still been on the top and when the shots were fired, they had dropped their gear preparing to advance. We were stretched out, just starting to move down.

As the squad advanced past and the medic worked on the wounded man's arm I studied the wounded man's face. I knew it hurt to be shot, but it was hard to imagine just how much. I guess from the conditioning movies had shown us that being shot was like scraping your knee. That obviously was not the case.

Initially the medic had not given the wounded man a shot of morphine, because the dust off would be coming soon, but later had second thoughts and decided to give him the hit of morphine. The wounded man was relaxing. The pain was subsiding or at least being disguised by the effect of the morphine.

While they were helping the wounded man back to the top of the hill to wait for the dust off, we heard a volley of shots ring out. The first were AK47, which was met with a barrage of M16 fire. In an instant it was over. The squad called up they had gotten him, as far as they could tell it was only one gook.

We pulled back to the top momentarily to regroup and start over, then started back down. As we came near the bottom the three platoons branched out and headed in separate directions.

Our platoon came across a bunker complex, but it was deserted. The complex looked as though it had been deserted for some time. We dropped a few grenades down for good measure. We were instructed to meet with the other two platoons in a flat land section.

The second platoon had discovered a supply depot. There were not any people, but there were ample supplies there, approximately two tons of rice. The rice had been stored in what could be construed as the basement of the biggest hooch in the complex.

The second platoon had also found weapons, ammunition, various gear and hospital supplies. There was a road leading into the complex as well.

Instead of blowing the complex like we would normally do, we called for an air strike to bomb the area. We packed up, moved out quickly and humped down the road to another open area. Helicopters were arriving to take us back to the rear for a couple of days. We had done our job for awhile. After resting we would move out again.

# ANOTHER FIREBASE

WE WERE GIVEN INSTRUCTIONS TO BUILD ANOTHER FIREBASE in an area of suspected enemy buildup. This firebase would stay. We were not going to tear it back down.

We landed in an open area that appeared to be a plateau. The sides were just a couple of feet off the rest of the flat land. We brought in the big guns again. This time we also had a mortar unit with us.

We started excavating the ground for the bunker sights. The sand bags and planks would be delivered the next day.

The lieutenant and I would share the same bunker. While in the rear, one of the clerks had reached his time to go back to the states. The lieutenant arranged to have the professor take over his duties as clerk. I became the senior RTO.

I also finally received my orders for promotion to Spec 4 (Specialist fourth class, equivalent to corporal). I had already started nagging the lieutenant to put in for my promotion to sergeant (buck sergeant, E5 class). I thought that if it took this long to make Spec 4, I wanted to be sure I made sergeant before I went back to the states. One thing about rank in a war zone, it becomes much easier to attain. One doesn't have to go through all of the bullshit to get promoted.

We had barely dug the hole by nightfall. We carved out sort of a pit behind the bunker for us to sleep in. The area around where our bunker was facing was wide open. There was absolutely no cover.

Around ten o'clock we were changing guard, with one man coming off and myself coming on. I was shuffling into position when we both heard it, the popping sound of something being shot through a tube. We both knew what it was right away. We were being mortared. I grabbed the horn yelling "incoming" into it, while the other guy started grabbing people to awaken them.

The first round exploded right in the middle of the big guns. Everyone was now awake. We could see the flash of the round being fired in the distance from our bunker. The gooks were out in front of us. Another set of rounds were fired and exploded in the middle of the big guns. The gooks really had us marked. They knew just where to fire. We were sitting ducks. Shrapnel was flying everywhere.

Our mortar unit tried to coordinate our mortars to the spot. Whether we did or not will never be known, but we succeeded in routing the gooks. After the mortar boys put ten rounds out there, the gooks stopped firing.

Those of us around the perimeter were unscathed, but the guys by the big guns had taken several casualties. There had been a piece of corrugated tin that one of the guys had used as a shelter to sleep under. The makeshift tent had taken a direct hit. The area where it was set up was now completely clear. No trace of the tin or the man sleeping in it.

Dust offs were brought in to take out the wounded.

In the morning, I went with a squad to see if we could find anything. We found the area where our shells hit and the spot where the gooks had set up their mortar tube. Our mortar guys had been right on target, but we didn't find a trace of anything. We went back to the firebase.

The higher ups sent out gunships to work the area around the firebase. The big guns were turned on the area and spent the day shelling.

The rest of us continued building the firebase. Barbed wire was brought out for us to string around the perimeter. As usual we set two layers out. The inner set of wire about fifty yards from the bunker and the second set of wire about five yards past that. Claymores were set all along the inner wire. This firebase was proving to be a lot better than the one we had previously built on the hill at Christmas time.

When the lieutenant was satisfied with our defense we laid back and relaxed. We went back to playing cards, not poker, just tonk.

With this area being wide open, there was not a chance to get out of the sun. The inside of the bunker, although it was shaded, felt more like an oven. There weren't any streams to cool off in either. We were all beginning to grow weary of this setup. The days were hot and long with the sun beating down on us all day.

The lieutenant felt that something was in the works. The construction of the firebase had stopped prematurely. It was true we had an adequate fortress of defense, but the firebase had room for improvement. We had also stopped receiving materials to build with and there had not been too much concern about further development.

The lower ranking guys, like myself, were content. We were not all that anxious to be construction engineers, especially in this heat. We continued our card playing and horsing around.

After about a week, the order came through to tear down the firebase and prepare to move to a new location. The word was they didn't care if we left the sandbags or wood planking, just tear it apart as best we could. The helicopters would be there the next day to take everything and everyone out.

We watched while the Chinook helicopters flew in to remove the big guns as they had done on that other firebase. Once that was accomplished, the helicopters took out the people supporting the big gun emplacement.

We had successfully torn down or destroyed the bunkers as best we could on such short notice. When the lieutenant expressed concern about the stuff we were leaving and such, I watched as the expression on his face changed. I heard a few mumbled yes sirs, and roger that. He handed me the handset back and shook his head as he walked away.

We were picked up in the afternoon and flown back to the rear. We would get another rest. We were going somewhere, but no one was saying where.

Our unit had done a lot of traveling over the last month. We had hopped from one place to another without too much hassle, but now we were just sitting around the rear waiting to move out to the next spot.

Most of us were curious as to why it was taking so long to move out and what the big secret was, although we didn't mind lying around the rear, eating regular food, sleeping all night without pulling a guard shift and smoking a little. Since we had been in the boonies most of the last two months we didn't get much chance to smoke. My once steady flow had all but dissipated. However my cigarette consumption was on the rise again.

Quite a few of us had been issued new or replacement gear for our rucks and we were instructed to turn in all of our current ammo for new ammo and new magazines. Of course we were ordered to "clean" our weapons. Typically all your gear was passed down or gathered from the re-supply bins when you joined the units. However, most of us already had new gear from the time we lost ours on Hill 474, so we didn't understand why we were getting new stuff again. Oh well, it gave us something to do.

On April 29, 1970, a force of American and South Vietnamese Soldiers launched an incursion into Cambodia. Now we knew what we were waiting and preparing for. We would be part of that force.

It was official. We received our orders. On May 5, 1970 we would be part of the next set of troops going into Cambodia for a two-week assignment. We would be airlifted into a province two hundred miles north of the Fishhook area about 75 miles north of Saigon.

A small piece of Cambodia jutted into Vietnam and was referred to as the "Parrot's Beak". This was the area we would be going in.

The NVA had been using Cambodia as a means to get into South Vietnam. The route enabled the NVA to reach strategic points in the south unmolested. Our forces would hopefully put a stop to that, or at least create a disruption, to that easy movement. That in effect was the game plan, as I understood it.

It had been suspected that GIs had crossed the border previously while chasing the enemy, but there was very little identification of what the border was and very hard to tell of you crossed or not. In any event, this time the United States had come right out stating its intentions.

## | 216 |  THE PROTECTED WILL NEVER KNOW

We all waited for the coming day with apprehension, not knowing what we would find there. It was a long wait.

# CAMBODIA

**WE WERE LOADED ONTO HELICOPTERS IN THE MORNING** and started out toward our objective. We were going in, in battalion strength. Over sixty helicopters were involved in our movement. It was the biggest combat assault any of us had been involved in.

After we crossed what appeared to be the border, the helicopters turned back and landed in a remote base camp. We received word that the lead gunship had been shot out of the air and we had turned back to look for a new route.

We had heard that it would be a little tougher in Cambodia, but having a gunship shot down was a bit much to cope with.

We landed in or near a place called Plei Shur Rang. It was a little town near the border. It had an airstrip, which is probably why we landed there. You just can't park sixty helicopters anywhere you want.

Once they decided on the new spot, we were airborne again. The whole battalion was brought down in the same area, spreading out far enough to avoid becoming one giant huddle of men. After everyone was on the ground, assignments were handed out.

Our A (alpha) company would stay in this spot to start constructing a firebase. The other companies would sweep out and report back here. We would work in company strength.

We all had quite a distance to cover to get back to the original area in which we were supposed to land. In fact, the maps we had did not cover the area we were in.

We found a nice little area to set up in for the night. It was a lot harder finding places with the whole company, but this one would be adequate.

No sooner had we settled in for the night when the trip flare popped. The claymores were blown. We heard screams of pain and shuffling about. We heard what appeared to be the sound of a man running into a tree, a dull sort of thud and moan. In the next instant it was quiet.

In the morning, again we found blood and a blood trail. We found something else this time, bicycle spokes. Whoever hit the trip flare was riding the bike. The claymore had heavily damaged the bicycle, but there was no trace of that either. The gooks had done a good clean up job again.

We spent the rest of the day and the next night in that spot running patrols to try to find anything.

Later that night we were listening over the horn to our B (bravo) Company's contact. Bravo Company had set up for the night and started taking mortar rounds on their perimeter. They were really in trouble.

Word came for us to go help. We were the closest to them. It was usually dangerous to move at night, but we were going to take the chance. We packed up and started moving toward bravo company's position.

The going was extremely difficult, but we kept heading for the light in the distance. We could see helicopters flying around giving support and taking the wounded out.

Some time later our company was halted. We had not been able to cover much ground in the dark, and whatever ground we could hope to cover would not get us there fast enough.

The contact subsided and hopefully Bravo Company would be left alone for the rest of the night. We looked for a place to set up. Everyone more or less dropped their gear in place and slept wherever they dropped.

In the morning we were ordered to go over to bravo company's location and pick up the pieces. I did not know they meant that literally.

Our company filed into what was left of bravo's perimeter and left others back to pull guard. Most of the guys I saw were still somewhat in a state of shock.

I went over to a guy I had come in country with. We had also gone through "P" training together. We sat down on a log next to each other. I lit a couple of cigarettes and passed one over to him. He took it subconsciously and after he had taken a couple of drags he looked at me and said, "I don't smoke". He finished the cigarette.

While others were trying to organize things, I looked around. The guy tapped me on the shoulder and pointed to a tree that had a huge section of bark torn out of it. At the base of the tree was a helmet, some odds and ends and a boot with something sticking out of it. He spoke softly.

"There had been a man, wounded, laying against that tree… the medic went over to help him… a round hit… that's what left…"

My eyes asked the question. His eyes answered it. We sat in silence.

## THE PROTECTED WILL NEVER KNOW

The company was ready to move out. Word was there were over thirty wounded taken out by helicopter the night before. There were at least eight dead accounted for. We were taking those bodies with us. The dead were not put on the helicopters. With that many wounded, the dead did not matter. The unit would not know the final outcome until later. Whatever it was, it definitely had been worse than anything else we had encountered. The NVA had sent their message to our bravo company.

As we prepared to move out, I took a last look at what was left of bravo's setup. There was a lot of blood and stuff around, scattered pieces of gear, spent shell casing and literally pieces to pick up. I had seen a lot of blood before, but never this much belonging to GIs.

We moved toward an open area we had passed on our way to bravo's location. The remaining guys from bravo company would be flown back to the firebase. Most of them were in a state of shock and were becoming trigger-happy. We could only imagine what was going through their minds and tolerated their random shooting up of trees and birds and anything else that made a noise on either side of our position.

We finally got what was left of bravo company on the helicopters along with the body bags. It appeared like the company had been reduced to a single platoon.

The next day, as we were working our way back, a helicopter flew into the area. The pilot radioed that he had some people for us. We were joined by two war correspondents. They wanted to do an article about the ground forces in Cambodia.

Our third platoon was humping the lead, with the second platoon following and ours, the first platoon, humping last. The correspondents were back with us and the CO had come back to join them. We had stopped for a rest while the correspondents interviewed us.

I was listening to the correspondents talk to the lieutenant and the CO when we heard a blast of gunfire. It was not at us but further up the column. Word came over the horn, our third platoon had made contact.

The CO had taken the horn from me and was getting an update on the action. I looked over at the lieutenant and he was standing by himself. The CO noticed too and raised his hand. The lieutenant pointed to a fallen tree a little bit in back of us. The correspondents had taken cover behind it. They lay there nervously looking around, as the rest of us hadn't moved.

I guess you do get a little hardened out here and you learn when to duck and when not to. It was easy to tell the gunfire was a safe enough distance away, so we didn't react. However, the two correspondents would have no way of knowing and had sought cover. Slowly they made their way out from behind the tree and rejoined the lieutenant, brushing themselves off as they did. The CO had his back to them as he listened to the update on the horn. I could see a big smile on his face.

The firefight was over as quickly as it had started. The third platoon spotted six NVA walking down a path and fired, killing five and capturing one. They also had a man hit, not bad, but he would have to be medivaced out. They were bringing the wounded man and the prisoner back to our location and the lieutenant was asked to send men to meet them to switch off.

The two correspondents were anxious to leave also. The CO calmly informed the two that they could go only if there was room, but the two were persistent in their view that they should be first.

## THE PROTECTED WILL NEVER KNOW

The battalion commander came on the horn asking for the CO. I intercepted the call and motioned for the CO to take it. After a few minutes of conversation, mostly the CO explaining the situation, he handed the handset back to me. The CO yelled over to the two correspondents that they would be picked up by the battalion commander's helicopter.

The next day, we moved into an area to wait for our alpha company. Alpha was linking up with us temporarily to coordinate our patrols. Alpha had come off the firebase to take over bravo's area.

Alpha had joined us around noon. While we were together we all broke for chow. The men in charge had finished the meeting and were preparing to move into their areas. Alpha was set up a short way off to our right. A few of us on the inside of the perimeter had started a poker game. We had not played in awhile.

The lieutenant came over to fill us in on the plans. Alpha would move out first, then we would move to another sector to start running more patrols. We would be here at least another couple of hours. Which meant there was plenty of time to get in a good game. We went back to playing.

We had just started a hand when all hell broke loose. We were being riddled with AK47 fire. Everyone hit the ground. As was customary in these cases, one of us would grab the pot and hold it until the shooting was over. It was not unusual for the gooks to fire a few rounds at us then disappear. It would be just enough to harass us. This time I grabbed the pot.

I started crawling toward my ruck. It was about fifty feet away, but it offered the best cover I could see. As I crawled along, I stopped to look and get my bearings. There were two M16s against a small tree. They had been crisscrossed and braced against each other to keep them upright. While I looked at the M16s, a burst of gunfire flew in, tearing the hand grips right off those rifles. I was showered with the debris.

I made a leaping dive for my ruck and landed right in front of it. I grabbed at it, trying to spin it around in the direction of the shooting as bullets ricocheted around me.

The fire that exploded inside my ankle first brought fear, then anger, then tremendous pain and finally a bizarre sense of relief.

I lay there motionless, even with the throbbing growing stronger in my ankle. I did not know what to do. I looked over at the lieutenant. He was crouched behind his ruck dodging the ricocheting bullets as well. I called over to him that I had been hit. He yelled back for me to lie still, that everything would be okay. In what seemed like hours, but probably were just a few seconds, the firing stopped.

The new sergeant was the first one to get to me. He was rolling me over on my back. I had made that last dive face down and the impact of the bullet hitting, had spun me to the side of my ruck.

As I turned over I braced myself up on my elbows to watch. The sergeant peeled off what was left of my new boot. The bullet had hit on the inside of my left ankle and exited through the top of my foot. The bullet had taken most of the tongue and shoelaces of the boot with it. The sergeant used his knife to cut away what was left of my sock.

I noticed a bubbling hole on the top of my foot. Even though the sergeant was trying to fit a bandage onto the top hole, it was not very effective at stopping the bleeding. The blood kept gushing out. By this time the medic had reached me and gave me a shot of morphine to combat the immense pain I was experiencing. Now I knew what all those other guys had gone through. The medic was able to get a series of wraps around my wound to basically stop the gushing blood, but you could see the bandage filling up with the red stuff.

I lay back against my ruck watching the rest of the activity around me. There had been others shot. I was not the only one wounded. From what the lieutenant could gather, the gooks or NVA had gotten between alpha company and us and opened up on both of us, hoping to get us to shoot at each other. We had not taken the bait and did not fire back. We knew how close we were to each other.

The medic came back to see how I was doing. While he changed my bandages he gave me a rundown on the rest of the wounded. One guy had been shot in the buttocks, leaving a hole the size of a half dollar. Another guy, a new guy, had been shot in the head. The new guy had joined us the day we were mortared on the firebase. Head wounds were permanent. They were usually carried out in body bags. I shook my head and sat there silently.

The guys came over to say goodbye and actually to start stripping my ruck. I would not need my gear anymore. My ankle was obviously broken. I would be going back to the world. Wasn't that what Jack had said? Jack? I hadn't thought about him in awhile. Christ, what day is this? Hell, what month is it? My mind was racing. I was trying to collect my thoughts. The morphine had put me in a relaxed state, but I had to know the answers. I called the lieutenant over.

"What is it, kid?"

"What is this?"

"What is what?"

"You know, day, month, where are we?"

"We are in Cambodia. Take it easy, kid…"

"I know where the fuck we are. I wanna know what time... not time, shit day, month, something for Christ sakes. What is it?"

"It's Tuesday, May 12..."

"Ah shit... I don't believe it. I'll be Goddamned. He was right. The son-of-a-bitch was right. Wait until I see that prick, I'll..."

"What the hell are you talking about, kid?"

"Jack. Jack said I would get it in May or June the latest and he was right. He was right that son-of-a..."

"Take it easy kid. Don't get worked up. You'll be outta here soon, just take it easy..."

"Oh, its cool sir, I'm alright, but you gotta do something for me, sir. You gotta promise, you gotta do it."

"What, anything, just name it."

"I gotta make sergeant. I can't go back to the world a spec 4 (specialist fourth class, use to be the rank of corporal), gotta be a five not a four. You promise? You gotta do it or I'll get fucked back there. You promise?"

"Yeah sure, kid. I'll do that for you. Okay now just relax. You'll be outta here soon."

I lay back and relaxed. I remembered I had scooped up the pot. I took it out of my pocket and was about to give it to somebody, but I decided against it. Instead I took out my wallet, placed the money in there. There had been sixty dollars in that bundle. I pulled out my IOUs that I had accumulated from others and gave those to the lieutenant to settle up any debts I had outstanding. I waited for the medivac helicopter.

Word came that the dust off was coming into the area. My radio was still next to me and as my last official act as an RTO I acknowledged their call. Someone else could take over. I tossed the handset to the ground. My job was finished.

They started gathering the wounded up to take out to an LZ that had been set up and secured. I got to my feet rather shaky leaning against one of the guys for support.

I was standing waiting my turn to move when I saw him, the new guy that had been shot in the head. Apparently the bullet had only grazed the side of his face and had not penetrated anywhere. I could see though by the amount of blood on his face and clothes why the medic had been skeptical. A feeling of relief washed over me. The new guy had not been killed. I hobbled, with the help of two guys, out to the LZ and waited for the helicopter to land.

The helicopter set down about four feet off the ground. With two guys boosting, the medic on board pulling and my right foot bracing, I climbed aboard and sprawled across the inner floor, resting on my elbow. When everyone else was loaded the helicopter took off.

We had barely taken off when the pilot headed back and started to land again. Panic seized control of my whole body. I could not understand why he was going back down.

The medic on board tapped me on the shoulder, pointing to the empty door-gunner's seat. Medivac helicopters did not have door gunners. At first I was confused, but then I understood. The medic wanted me to move over there. I crawled. It was easier than trying to stand again. Another guy was sent to join me on the door gunner's seat. Once we landed again I understood why.

Two more wounded men were loaded on. They were from alpha company and were further back from the LZ. One man was not moving. He did not look good at all. After a few minutes the medic slid his index finger across his neck then covered the man's face with a blanket. We flew to a field station.

The wind from the flight had blown my pant leg about and

had succeeded in saturating my other pant leg in blood giving the indication both legs were hit.

Upon arrival at the field hospital we were immediately rushed inside for evaluation, before being sent on to the hospital. My wound was redressed and an air cast was placed on my leg. While I lay there I saw two guys bring in the dead guy on a stretcher, but the doctor motioned for them to take him out saying something about only wounded in here. Eventually, I was placed on a stretcher and loaded onto another helicopter for transport.

At the hospital I was brought inside and placed on a table. Two nurses and a doctor were starting to work on me.

The nurses each grabbed a pant leg cutting it straight up and right through the waistband. They peeled the front down, propping me up slightly, while they pulled it free of my body. An orderly had helped me remove the shirt. In an instant I was lying naked.

A washcloth had been placed over me to protect my modesty, but the excitement of seeing two American women caused a reaction within me. The reaction was quickly taken care of by the nurse on my right. She placed her hand on the bulge in the washcloth and pushed it back down. All she said was it happens all the time and smiled.

My thoughts were replaced by the activity happening around me. Each nurse was puncturing an arm with needles. When they finished, I had blood going into one arm and an intravenous feeding bottle in the other. The doctor had been working on my ankle. When he was satisfied with his work he instructed the nurses to prepare me for surgery and then he left.

An orderly brought a sheet over to cover me while I waited my turn for the operating room. One of the nurses let me suck on a wet washcloth to quench my thirst, but because of my going into surgery, she would not let me have anything else.

Finally my turn came. I was wheeled into the operating room. The sheet was removed and a green sheet was placed over me. Someone told me to relax and to start counting backwards as soon as the mask was in place.

"One hundred, ninety-nine, ninety-eight, ninety-seven, ninety…"

## THE MORNING AFTER

**I** **AWOKE ONCE DURING THE NIGHT.** I guess it was night. Actually my mind brought me back to consciousness to let me know it was time to throw up. I had been given gas in the operating room and gas always caused a violent reaction in me. Even when I would go to the dentist to have a tooth pulled and would be given gas, I would throw up. This time would not be any different.

I felt the urge growing stronger as I lay there. I called to a man in white standing nearby to bring me something to heave into. He brought me a little dish. I looked at him as if to explain, but the urge got the better of me and I let the first pass flow. The vomit immediately filled the dish and proceeded to splash onto the floor. Before I had a chance to recover the second pass was flowing. The dish was completely useless so I let the next flow drop straight to the floor. With more on the floor than in the dish, I let the dish plummet to the floor with the rest. I lay back and passed out again. The mess on the floor was not my problem.

Later, I was vaguely aware of being transferred from one bed to another and found out that I had been moved from recovery to a ward.

# | 230 |   THE PROTECTED WILL NEVER KNOW

I awoke suddenly, confused at first, not knowing where or why I was there. Slowly I looked around inquisitively, trying to remember everything that had happened.

The memory of the day before flashed through my mind, everything came back to me all at once. I tore the sheet off my body and pulled the blue hospital gown away as well.

The glowing white cast engulfing my left leg heavily accented the naked deeply tanned flesh of my body. I continued to stare at the sight curiously. I wasn't quite sure why the whole left leg was in a cast as I had been shot in the ankle. The cast extended from just below my toes right up to my pubic hair. The fact that I could see my toes was quite encouraging. It meant I was still intact.

A nurse came over to check on me. She helped rearrange the covers and noticed that the IV bottle stuck in my right arm had bloated the arm. She removed the IV saying that I should be able to have some solid food today.

"Today? What day is it?"

The nurse assured me it was only the day after surgery, I had not lost any days. She left, but came back with a pain pill.

I lay back and relaxed. The guy in the bed next to me was full of tubes. He hadn't moved since the first time I saw him. There were only two others in the ward with me and they were both from our bravo company.

The orderlies had put us in here because the other wards were full. This ward had been closed previously and had a very musty smell. We had been moved in during the night from the recovery room or at least I had. I wasn't quite sure about the others.

Later that morning an officer came to my bed and awarded my Purple Heart by pinning it to my pajama top.

*The following awards are announced.*

*Meyer, Donald SP4 D Co 3/506 101ˢᵗ AB*

**Awarded:** *Purple Heart – First Award*

**Date of Service:** *12 May 1970*

**Theater:** *Republic of Vietnam*

**Authority:** *By direction of the President UP of AR 672-5-1 and USARV Msg AVGP-D 08713,* **Subject:** *Award of the Purple Heart, dtd 4 Mar 66*

**Reason:** *Wounds received in action*

I lay there for awhile holding the ribbon of my new medal, envisioning the array of fruit salad (medals & ribbons) I could now wear on my uniform. In addition to the Purple Heart, I had also been awarded the Army Commendation Medal with "V" Device for Valor, the Air Medal and of course the Vietnam Service and Vietnam Campaign Medal. In addition I had been awarded a Combat Infantry Badge. Not a bad haul for my first year in the army.

After a bit, I removed the ribbon and put it back into the box for safekeeping. I folded the orders and stuck them in as well.

A girl from the Red Cross came around asking if anyone wanted to write a letter home. I called her over, I certainly did. She offered to write the letter for me, but I declined. I thought it was bad enough writing home on Red Cross stationary without having someone else's handwriting.

# | 232 |  THE PROTECTED WILL NEVER KNOW

I composed the letters as casually as I could, but there just wasn't an easy way to say that I had been shot, wounded in action, the whole Purple Heart bit. I decided to be blunt and lay it on the line.

Dear Mom,

Nothing new, got shot yesterday, everything else about the same...

How else can you say it? I wrote three letters, one to my mother, one to my girl and one to the gang at the office. I mailed the first and last. I decided to let my mother explain to my girl. I had hoped it would be easier to hear it instead of read it.

I had been shot on Tuesday, Wednesday I laid around regrouping my thoughts and fighting pain. The trauma of the gunshot, the broken ankle and the effects of surgery were all vying for position on my throbbing ankle. On Thursday, I had become stir crazy and made a gallant attempt to see a movie. Friday, I left for Cam Rahn Bay by helicopter to prepare for my flight to Japan and another hospital. The first stop on my journey home.

Several of us wounded would spend the night in Cam Rahn Bay, before being flown to Japan. This time the ward was crowded.

We also had to switch our money back to American greenbacks. The memory of the first switch back in Bien Hoa flashed through my mind as I handed over my MPC. I had over one hundred dollars, mostly from that last poker game. The clerk making the switch could not understand what was so funny, but there was no way I could explain the story to him. It was a private joke.

I had a tag affixed to my chest with the initials GSWLA/RVN-C. The tag was pretty much like you would find attached to a piece of furniture or an appliance, cardboard with a wire running through the hole twisted together, hung from the buttonhole of my blue hospital gown. I notice others had tags, but some did not.

As I lay there I tried to make out what the initials meant, but all I could get was RVN, Republic of Vietnam, the rest were meaningless. Of course the army used initials and shortened words for everything and half the time you could work through it, but this was an Air Force tag. I asked the nurse, but she said I'd have to ask the Air Force since it was their tag. I flagged down an orderly passing by, but he said the same thing so I asked him if he could get an Air Force guy. The orderly mumbled something as he left, but I didn't hear.

A little while later a clerk wearing Air Force fatigues walked up to my bed, tuning the tag so he could read it.

"What do you need to know?"

"What do the initials mean?"

"Means you are not ambulatory, you'll have to be carried to the plane."

"How the hell do you get that from those initials?"

"The initials describe your condition. Only the non-walkers have tags on them. All of the rest of your guys will have to be loaded on the plane."

"Loaded? What are we, baggage?"

"Yeah."

"You're serious?"

"Yeah, if you can't walk we'll have to load you in the back cargo section."

"Well shit…"

"Slow down solider, don't get your nuts in a tangle. It's SOP."

"Alright but what do the initials mean… GSWLA/RVN-C?

"Gun shot wound left ankle/Republic of Vietnam—Cambodia"

"Really…? That's all…? Hey, thanks…"

The Air Force clerk left my bed and the ward as quickly as he had come.

So I was nothing more than cargo? That was exactly how I was loaded onto the Air Force C-141 jet. I was the last one placed on the plane right in the back next to the cargo hatch. On a C-141 the whole back section opens up to load cargo on. I was right next to this section.

While I was waiting for the plane to finish loading I had a clear view out the back cargo hatch. I saw a freedom bird (commercial airliner) land and taxi to the hanger. I watched as the new "cherry boys" disembarked and filed into the terminal.

As I was watching, a group of American stewardesses walked toward our plane. They came aboard, spoke with us and visited. From my position in the back and the fact that I was about a foot off the floor, I was able to get a unique view of those stewardesses. One of the girls added to my pleasure by kneeling down to speak to me. Although it was obvious to her where my eyes were focused, she made no attempt to rearrange her skirt or shift her position.

I did make eye contact with her eventually and her lovely face and her beautiful smile further enhanced my already frantic condition. She bent over, patted my cheek and kissed my forehead. In the next instant she was gone. In fact, the stewardesses were all gone. I lay there trying to rearrange my body into a comfortable condition. It was useless, I would just have to wait it out.

Before long we were in the air, on our way to Japan. The flight was uneventful and we landed a short time later. We were then transported to various wards depending on our conditions. Most of us would stay in Japan for awhile, either for further recovery or additional surgery. In my case it was additional surgery. The doctors were going to operate on my ankle again.

We had arrived in Japan on Saturday and the doctors operated on Tuesday. That morning, I was informed that I would be going in. Actually the nurse told me the night before not to eat or drink after midnight, because I would probably go in the morning. Of course I woke up starving, but they would not give me anything until they were sure. I had also asked for a urinal to do my morning duty, but assuming, I guess, that I had asked for something else, they ignored me. Next thing I know I was being wheeled into surgery.

I received a spinal tap as they called it, which caused the lower half of my body to go numb. I watched as the doctor cut the old cast off. I continued to watch while the doctors went to work on my ankle. I wasn't sure what they wanted to do, but I didn't have much choice in the matter.

The operation took about an hour. I was in the recovery room at nine-thirty in the morning. The fellow who shared the bed next to me in the ward was also in there. He had a similar wound, also having his foot shot up. We both had received spinals and coincidentally we both still had to urinate, but no one would bring us a urinal explaining that we should wait until we got back to our beds.

We were not brought back to our beds until ten-thirty, a full hour later. While we were being placed back on our own bed we were instructed to lay perfectly flat, because, by getting up after we had a spinal tap we would get a tremendous headache. They even took our pillows away. However, the orderly did bring each of us a urinal.

Although the urge to urinate was stronger than ever it would just not happen, especially lying flat on our backs. After several different attempts, the guy next to me said he was getting up. He didn't really get up, but sort of manipulated into a more angled position than perfectly flat. Still nothing. Another hour passed.

I tried everything I could think of to get it to work, but all attempts failed. I could go so bad I could taste it, but nothing would make it flow. My eyes were beginning to float. We tried cracking jokes to each other, hoping that by laughing we could get something going. We hit every piss joke we could think of, still nothing.

The orderlies brought lunch around at noon. The site of liquids made it worse. I laid my hand on the hot lid covering the meal hoping the heat from the metal lid would warm my hand sufficiently to cause a reaction, but as the guy reminded me, it only worked with warm water. Another failure.

I motioned to one of the orderlies covering the floor and let him know my problem. He gave me a leering smile and said:

"If you don't pee in the next half hour, I'll run a tube up you. You gotta do it before your bladder bursts. You or me, makes no difference."

That did it, no matter what it took, I was going to do it myself. I got up and the guy next to me asked what I was doing. My only reply was, that I was going to piss. He watched for a minute while I struggled to get into position, and then decided to join me. I was almost sitting up, but nothing happened.

I heard a moan and looked over. He had done it. He was finally going. I asked how he did it. He looked over. "Fart, rip a fart, that will do it." I followed his advice. It happened, I was going. I was finally going. Tears were streaming down my cheeks. Peeing was both painful and pleasurable at the same time. The sensation was tremendous. My whole body shook from relief. I never knew urinating could be so exciting.

I kept maneuvering the urinal so as not to overflow it. I kept filling it. I didn't think it would ever stop. I practically had the urinal upright trying to catch the flow. Finally the last of it drained out. I let it lay there in the spout, making sure every last drop had been drained. When I finally did pull it out, one last drop rolled off the tip and onto my leg as if in defiance. I had to use two hands to lift the urinal onto the table next to my bed. I had never seen one that full. I laid back waiting for the headache to come, but it never did.

I spent the next week lying in bed, because I had to keep the leg elevated. It was boring as hell, but what could I do. I played cards, checkers and chess a lot to pass the time.

Finally I was able to get out of bed for short periods of time. I made it over to the PX (post exchange) and bought a stereo, which I had shipped home. I also got to see a couple of movies. At least it gave me something to do at night.

Word came down that we were being shipped back home, but then we were told we would have to wait until after the Memorial Day weekend. We had to leave Tuesday instead of Friday. Eventually we were transported to another base for preparation for the long flight home. We spent the night there.

# GOING HOME

**A GROUP OF US WERE LOADED** (I was still traveling as baggage) onto an Air Force C-141 transport for the flight back to the states. The flight departed at ten-thirty AM Wednesday. After flying for what seemed like forever we arrived in Alaska at eleven-thirty PM Tuesday. The night before we left Japan. The plane stayed on the ground long enough to pick up more passengers.

In a short while we were airborne on the last leg of our flight home. As is customary, the army tries to place you in a hospital close to your home. I was from Chicago, consequently I had been assigned to Great Lakes Naval Training Center for further hospitalization. Great Lakes is about forty-five minutes north of Chicago in Illinois. Three guys from Michigan and one from Indiana were also with me. I was lucky being that close, it would be easier for me to go home. I did not know if I could do that or not, but the thought of going home really boosted my morale.

We landed, bused and processed into Great Lakes by approximately three-thirty PM Wednesday, the same day we left Japan. All the flying, which seemed like weeks, only actually covered five hours of time. Talk about jet lag. Most of us that made the trip could not condition our minds and bodies to the time difference. We were awake at night and slept during the day.

I spent that first night trying to reach someone at home. My usual flawless memory of phone numbers had failed me miserably. I could not remember a single number.

I had been assigned to a contagious ward because of my entry wound infecting on the flight back to the states. There were not any phones in the ward itself, but there were a couple of pay phones outside the ward that all of us had gathered around. I was using a wheel chair to get around and it was difficult to reach the phone from a sitting position, but I was able to pull the receiver down and hit the operator button for assistance.

I explained my dilemma to the operator and she was quite helpful in finding and reaching a number for me. I was able to reach my mother, but had to tell her not to come visit. Due to the infectious nature of my wound, it was not advisable to have visitors. Perhaps by the weekend she would be able to visit, the nurse had said. At least my mom knew I was back in the states.

On Friday, I was given permission to go home, with certain restrictions. Someone had to constantly clean my wound. That meant running a large size q-tip into the hole in my ankle until the wound was clean. In other words once the yellow green combination turned blood red. My mother is the type who gets nauseated at the sight of scratches. My chances didn't look too good. However, brave soul that she was she agreed to do it, and I could go home. My mom and soon to be step-dad would pick me up at noon.

It suddenly dawned on me that I didn't have any clothes to wear home. A Red Cross girl was walking through the ward asking if anyone needed anything. I called her over and asked if she would go to the PX to buy me some clothes. At first she refused, claiming she had never bought men's clothes and would probably buy the wrong things. After I convinced her that she would be the cause of my not being able to go home after spending six months in Vietnam she finally relented and agreed to do it for me. I gave her money and sizes and sent her on her way.

I contacted the head nurse about getting a pair of crutches to use. The head nurse informed me that I had not been authorized to have crutches. She indicated that I could use a wheelchair, but I would have to order one and this late on a Friday, the wheelchairs were probably all gone by now. When I convinced her that I was going home even if I had to crawl out of there, she somehow managed to round up a wheelchair for me. It was amazing what people will do when they are properly persuaded.

The Red Cross girl returned with my clothes. A shirt, a pair of slacks, a package of undershorts, briefs not boxers, and a pair of black socks. That would be adequate I thought.

After the operation in Japan, I had been fitted with a short cast, from my toes to just below my knee. Upon my arrival at Great Lakes Naval Hospital, I was again put back into a long cast, from my toes to my pubic hair. I got the shirt on okay, but my progress stopped there. I sat there frustrated. There was just no way for me to maneuver around that cast.

I had become fast friends with the guy in the bed to my right. After he watched me suffer awhile, he offered to help. He put the under shorts on the cast leg pushing it up far enough for me to reach. He then started working the slacks on. Obviously the slacks would have to be cut. In the sixties everyone still wore those very tight tapered pant legs. He cut the leg to about five inches below the crotch, before he was able to get the slacks up to my waist. I put the sock on the one foot. No shoes. I did not have any shoes. I was about to send someone after a pair of black shoes when I remembered I would be in a wheelchair. I breathed a sigh of relief.

The nurse came over with a bag of medical supplies for me to take home to clean my wound.

The doctors had cut a hole in the side of the cast, which they called a window to be able to get to the infected wound. Periodically, the doctors came around, removed the protective covering of an ace bandage, then the gauze pad stuck in the hole to expose the wound. Next a long stemmed q-tip was stuck into the wounded area and the wound was scraped until it showed blood red again, removing the yellow green mess that formed in the wound. The probing q-tip would often hit an exposed or severed nerve ending in my ankle area causing me to jump or pull back so usually the nurse held onto my leg while the doctor probed and I held onto my pillow or the bar on the bed for support. Once the wound was cleaned the nurse would repack and rewrap the window covering the wound back up, while the doctors moved on to the next patient.

This is what my mother would have to do at least three times a day while I was home. The bag contained a supply of those items. As I peeked inside the bag, I just could not visualize my mother duplicating the procedure, nor my sitting still while she did it. Didn't matter I was going home, we'd work out the details later.

My mom and step-dad picked me up around twelve-thirty in the afternoon on Friday. We all went over to a building to sign out. No matter what, the army still had rules. In effect I had a three-day pass, so I didn't have to be back until Monday, but my parents would bring me back Sunday night. So home here I come. We were home in an hour.

The wheelchair had not been any trouble when we closed it, but opening it was another story. I decided to hop into the house. A friend I grew up with lent me a set of crutches. I used the crutches to get around the rest of the weekend.

After I got in the house and settled down for a minute my stepfather handed me a telegram that came that morning to my mother about me.

> 1970 June 5, AM 9:40
> "Your Son, SP4 Donald P. Meyer USA has arrived this hospital. He has been hospitalized as a result of broken left ankle and his condition is considered satisfactory. His medical officer does not feel that your presence is required from a medical standpoint, although you may visit him anytime on your first visit and from 2:00 pm to 4:00 pm and from 7:00 pm to 9:00 pm daily thereafter. He is able to write at this time. His mailing address is ward 3S Naval Hospital, Great Lakes Illinois 60088. You are assured he is receiving the best medical care and you will be advised of any significant change in his condition. Signed Captain MC USN Commanding Officer"

Fortunately, I had already spoken to my mom and since my stepdad picked her up from work on his way to get me, I was home before she saw the telegram. (I still have the original telegram.) Oh well, at least somebody told her where I was and sort of how I was doing.

I had my mother call my girl friend and tell her there was something at the house for her so she should stop by after work to pick it up. I did not tell her that something was me. I had to get the last laugh.

## THE PROTECTED WILL NEVER KNOW

I was finally home. I had also completed my first year in the army. After a rigorous four months of training, a three-week leave, six months in Vietnam and a month so far in the hospital I was ready for anything. I was a hardened veteran now. I could worry about that later, I was home now, with an indefinite stay ahead of me at Great Lakes. I could lay back and enjoy life for awhile.

It was in this state of euphoria that I innocently left the house to visit friends I had not seen in awhile. My progress was slow. I had been lying around too much and navigating those crutches was awkward. The walk was tiring.

As I paused to rest at the corner one of the ladies in the neighborhood walked up asking what had happened to me. I was barely able to get out the words out that I was shot… when she quickly responded, "See, you son-of-a-bitch, that's what you get for hanging on the corners all the time, serves you right." She walked away before I could explain.

I stood there leaning on my crutches, watching her walk away. I kept thinking about what she said. I wondered what happened to the place while I was gone. Was the Vietnam War that far out of everyone's mind?

Will the real enemy please stand up?

Welcome home soldier, I heard…

Every bit of ecstasy, every bit of my euphoria was instantly drained away. No one knew, no one cared, why should I? What was it all about? Did I miss something? I was only gone six months. Did the world change that much?

Will the real enemy please stand up?

Will the real enemy please stand…

Will the real enemy please…

Will the real enemy…

Will the real…
Will the…
Will …
I wished I could stand…

# EPILOGUE

Another question that was often asked is what happened next. While I focused this memoir on the specific time I spent in Vietnam/Cambodia and not my overall military experience, I thought it appropriate to "finish the story."

As I mentioned in the Going Home chapter, I was reassigned to Great Lakes Naval Training Center just North of Chicago and was able to go home on occasion. It was there I had another operation to repair the damage to my left ankle, along with several cast changes. I eventually spent the summer there before being reassigned to Ft. Hood Texas.

Upon reaching Ft Hood, I was assigned to an infantry battalion (Patton's old division — 2nd Armored Division). However, as I was just learning how to walk again, I was put on restricted duty and spent most of my days in physical therapy. I went through a series of assignments from CQ (Charge of Quarters) to finally landing a job in the A&D (Admissions & Dispositions) section of the hospital — hell I was there just about every day anyway.

I was put through a battery of tests and exercises to determine the degree of disability over a period of about six months and eventually I went before the medical board that determined my disability was permanent and started the process to discharge me, which they finally did a year after arriving there with permanent disability status.

I went home—and got on with my life.

[Note: In the last couple of years I've had the opportunity to reconnect with a few of the guys in my platoon—Rags, Doc, The Professor, LT, the Captain and a few others by email. In 2010 we had a 40th anniversary reunion of the Cambodia incursion that enabled me to actually meet up with a couple of these guys as well.]

www.ingramcontent.com/pod-product-compliance
Lightning Source LLC
LaVergne TN
LVHW011416080426
835512LV00005B/79